An Introduction to Animal Cytogenetics

An Introduction to Animal Cytogenetics

H.C. Macgregor

Professor of Zoology
University of Leicester

CHAPMAN & HALL

London · Glasgow · New York · Tokyo · Melbourne · Madras

Published by Chapman & Hall, 2–6 Boundary Row, London SE1 8HN

Chapman & Hall, 2–6 Boundary Row, London SE1 8HN, UK

Blackie Academic & Professional, Wester Cleddens Road, Bishopbriggs, Glasgow G64 2NZ, UK

Chapman & Hall Inc., 29 West 35th Street, New York NY10001, USA

Chapman & Hall Japan, Thomson Publishing Japan, Hirakawacho Nemoto Building, 6F, 1-7-11 Hirakawa-cho, Chiyoda-ku, Tokyo 102, Japan

Chapman & Hall Australia, Thomas Nelson Australia, 102 Dodds Street, South Melbourne, Victoria 3205, Australia

Chapman & Hall India, R. Seshadri, 32 Second Main Road, CIT East, Madras 600 035, India

First edition 1993

© 1993 H.C. Macgregor

Typeset in 10/12 Times by EXPO Holdings, Malaysia
Printed in Great Britain by The Alden Press, Oxford.

ISBN 0 412 54600 0

A catalogue record for this book is available from the British Library

Library of Congress Cataloging-in-Publication data
Macgregor, Herbert C.
    An introduction to animal cytogenetics/H.C. Macgregor. – 1st ed.
        p. cm.
    Includes index.
    ISBN 0–412–54600–0
    1. Cytogenetics.   2. Chromosomes.   I. Title.
QH441. 5. M33 1993
591. 87' 322–dc 20   93–13470
                                                                        CIP

∞ Printed on permanent acid-free text paper, manufactured in accordance with the proposed ANSI/NISO Z 39.48-199X and ANSI Z 39.48-1984

Contents

Preface

There were three reasons for writing this book. First, it marks the 25th anniversary of the first occasion on which I lectured to first-year undergraduates about meiosis. Second, it is written at a time when some very exciting things are happening in the field of chromosome science. Third, its publication is timed to underline the crucial importance of understanding the supramolecular, the visible things that go on in cells, so as to provide the shapes that are to be coloured in with paints from our molecular paintbox.

The book is directed at both teacher and student. It is not a major text. It is written by a zoologist and unashamedly lacks any consideration of plants. It begins almost at the beginning, but it depends on a basic understanding of animal cells, animal diversity and the evolutionary process. By far the most important aspect of the entire book is contained in Chapter 3, which is about meiosis. In my experience, young people do have some difficulty in coming to terms with the intricacies of the meiotic process, and I have found that a good way to bring it alive and expose the astonishing command that it exerts over living organisms is to study it in depth, to deal with the formalities and then to concentrate on what we simply have not yet managed to discover about this astonishing sequence of chromosomal behaviour. Understanding meiosis provides the confidence to tackle things like polyploidy, aneuploidy, sex and evolution. So I have given meiosis, its knowns and its unknowns, the fullest treatment, almost to the frontiers of modern biology, that is consonant with an 'introductory' text.

In every chapter there are questions, many of them left unanswered. They are of three kinds. There are questions that represent the boundary of scientific knowledge in 1993. Watch these spaces, for they will sooner or later be occupied by new discoveries, and it is entirely possible that you will be one of the discoverers. Make no mistake: chromosome science is one of the great uncharted oceans of modern biology. There is plenty of room for adventurers!

Then there are smaller questions that you could probably answer quite easily if you were to design some simple experiments and had access to some basic equipment. These are the kind of matters that help to develop our intimacy with living systems and also help hard-pressed teachers to come up with doable mini-projects for eager students.

Perhaps most importantly, there are lots of little questions that are posed as part of a drive to inculcate the habit of minute-by-minute curiosity about the visual images that confront us when we look down a microscope. How big? How many? How fast? How frequent? Only by asking and taking the trouble to record and measure can we bring these images to life.

This is, inevitably, a personal book, reflecting the particular experience and interests of an author who has spent his entire professional life working with animal chromosomes. Molecular biology is not a major theme: molecular matters are expertly and lucidly explained and discussed in a wide range of current publications: they are mentioned here only where it seems reasonable to pursue a certain cytological phenomenon right down to its molecular roots. Most of all I hope that this book will be of real value to students who wish to learn about chromosomes and their role in evolution and development. Their next steps towards the frontiers of knowledge in animal cytogenetics would probably be the timeless and immensely valuable works of the late Michael White (*Animal Cytology and Evolution*, 1973, and *Modes of Speciation*, 1978), or John and Miklos's more recent and eminently readable book on *The Eukaryote Genome in Development and Evolution*.

Chapter 11 offers a quick flash of modern clinical cytogenetics. The strategy here was not to compress this enormous, complex and fast-moving field of science, but rather to outline the kind of things that actually happen in an average clinical laboratory in the 1990s. It has been my experience that such an approach, coupled with a brief visit to a busy regional laboratory, is enough to help students decide whether or not they wish to become involved on a professional level. In this regard I am particularly indebted to my colleagues Dr Michael Schmid of the Human Genetics Department at Wurzburg, Germany, and Dr David Duckett, Director of the Leicester Area Health Authority's Genetics Unit, for advising me on the current clinical role of their departments.

I was encouraged by a number of colleagues to include the Appendix as a step-by-step guide to four laboratory teaching sessions in animal cytology. Each of the protocols has been used over and over again with consistent success and each is considered to be useful in three important respects. First, they are all inexpensive. Only the human cytogenetics exercise requires significant expenditure on materials and reagents, and even that can be operated successfully with preparations supplied on loan from a local clinical service laboratory, thereby eliminating the need for cell culture. Second, they are all busy and usually rewarding for students. Each person ends a session with something that they have individually made and can take away. Third, they are all based on fresh material, offering opportunities to see the tissue of origin, the living cells and then the fixed and stained chromosomes from these cells.

I owe special thanks to all my colleagues who have helped me to produce this book: to my senior colleagues and close friends, Mick Callan (University of St

Andrews), Joe Gall (Carnegie Institution, Baltimore) and Jim Kezer (University of Oregon), to Stan Sessions (Hartwick College, NY), Jennifer Varley (Christie Hospital, Manchester) and Alma Swan (Leicester) and very specially to Dr Irina Solovei (University of St Petersburg), who provided all the original drawings for the book based upon our consensus view of the current state of knowledge with respect to each of the structures that are represented. I also thank Penny Butler and Lesley Barnett for having so capably attended to the administration of my department and the servicing of my research and teaching programmes that I was able to devote time to this particular task.

H.C.M., Leicester
December 1992

Chromosomes in mitosis

This book is about chromosomes, their form, their behaviour, their inactivity and activity in relation to replication and transcription, and their evolution. It will begin with a simple consideration of chromosomes in mitosis, not a description of mitosis but an introduction to the concept of chromosome number and shape, karyotype and C-value as true phenotypic characters that have evolved through natural selection just like all the other features that we use when we describe a species of animal or plant. The book will end with discussion of some of the more challenging questions regarding the molecular basis of chromosome structure, function, behaviour and evolution. For the most part we shall be dealing with animal systems, although examples from plants and microbes will be used wherever it seems appropriate. Certain kinds of animal will feature more often than others, usually because they are particularly well suited, for one reason or another, for looking at chromosomes. For example, there will be a lot about amphibians because some of them have the largest chromosomes in the world. There will be a lot about mammals, including man, because their chromosomes show interesting variations from species to species, and good technologies have been developed for mammalian tissue culture and the study of human chromosomes. There will be a lot about insects, mainly because they present us with some of the most bizarre chromosomal situations that are known, and also because *Drosophila* happens to be an insect.

In the course of learning about animal cytology, two important things should become apparent. First, cytology is a science of seeing. We are at all times dealing with visible phenomena and we shall be busy for much of our time with the analysis of form and with explanations of changes of form in molecular terms. Time and time again we shall see examples of situations in which the end product of an experiment is a picture rather than a number or a dataset. Not surprisingly, then, cytologists tend to be good with microscopes and cameras and to have a strong sense of aesthetics. Secondly, the study of animal cytology draws heavily upon evidence that has come from technologically simple studies of exceptional natural situations, on the simple principle that, as often as not, the exception proves the rule. For the most part, cytologists search for and exploit natural variation in their quest for clues to the directions, limits and possible mechanisms of events that take place in animal cells. A good appreciation of animal diversity is therefore useful for this particular branch of science. In general, zoologists make good cytologists.

THE CELL CYCLE

Let us begin with some of the basic language of cytology and a particular focus on normal cell division. Mitosis, or the ability to produce two cells from one (i.e. to divide equationally), is arguably the most important function in a multi-cellular organism. If mitosis stops then the organism will very quickly grow old, wear away and die. Mitosis is the principal source of growth and the only mechanism for replacement. The frequency with which cells divide in an average, young, healthy, human adult is of the order of 20 million mitoses per second, the vast majority of these occurring in the bone marrow, the skin and the lining of the intestine, with other divisions taking place in the lymph nodes, the spleen, the hair follicles and, of course, in the vicinity of wounds. The re-markable thing is that, even at this truly phenomenal frequency, mitosis is a process that rarely goes wrong. When it does fail, natural selection is generally unforgiving and the consequences are inviable cells, except of course in the case of cancer.

The **cell cycle** consists of four stages. In two of these, the cells are presumed to be Growing in size (**G1** and **G2**). At another stage they are Synthesizing DNA for the replication of their chromosomes (the **S-phase**). In the fourth stage they are dividing, mitosis. Mitosis is conventionally subdivided into **prophase** (**P**), **metaphase** (**M**), **anaphase** (**A**) and **telophase** (**T**). Cell cycles can therefore be described as:

$$\underbrace{G1 \longrightarrow S \longrightarrow G2 \longrightarrow}_{\text{Interphase}} PMAT \underbrace{\longrightarrow G1 \longrightarrow S \longrightarrow G2 \longrightarrow}_{\text{Interphase}} PMAT$$

and so on. It is probably most convenient to begin our study of mitosis with a cell that contains the full **diploid** number of chromosomes embodying the **2C** amount of DNA that is characteristic for the organism under study. This is a cell that we would normally suppose to be in G1, thereby implying that it had not yet replicated its chromosomes, it was in interphase, and sooner or later it would begin an S-phase during which DNA would be synthesized and the chromo-somes would be replicated semiconservatively.

At this point we can ask our first question, albeit a preliminary one, since we shall not take the trouble to answer it. What initiates mitosis? Can we regard the initiation of mitosis as a state of differentiation of the cell in the sense that a cell can either be expected to, or would normally not be expected to, undergo mitosis. The cells of a plant meristem or those of a frog embryo regularly divide.

So do the cells of our bone marrow, our skin or our intestinal epithelium. On the other hand, the erythrocytes of a chicken or a frog, even though they have all the necessary components and equipment, would not normally be expected to divide, nor would muscle cells, spermatozoa or unfertilized eggs. Yet all these cells types, and indeed all cells that have nuclei, no matter how structurally differentiated they may be, can probably be induced to begin the processes that lead on to cell division.

Essentially what we are talking about when we speak of the initiation of cell division is the initiation of DNA synthesis, since in every case, except only for the second meiotic division in the formation of gametes, DNA synthesis is a prerequisite for cell division. However, it is important to note that the synthesis of DNA is not necessarily followed by cell division; hence we have polyploid cells, polytene chromosomes and amplified genes, as we shall see later.

CHROMOSOME NUMBERS

The second stage of mitosis commences with the condensation of chromosomes. This is normally followed by spindle formation and a movement of chromosomes that leads to the orderly separation of the products of the chromosome duplication that took place during the preceding interphase and the distribution of these products to opposite parts of the cell in readiness for cell division.

Chromosomes! We can stop right there and ask – why chromosomes? Why does a cell nucleus contain its genetic material in a definite and highly specific number of blocks rather than in one large piece or as individual gene-sized particles? There is no direct answer to this question, but I hope the sense of the situation will emerge as we go along.

In this context it is appropriate to ask about chromosome numbers in animals. The range is wide. At one end of the scale we have a fully authenticated haploid chromosome number of 2 in the iceryine coccid from Mexico, *Steatococcus tuberculatus*, an insect that is related to the cottony cushion scale that used to trouble Californian citrus farmers and the mealy bugs that live on old potatoes and corn. At the other extreme, we have the small blue lycaenid butterfly, *Lysandra atlantica*, with a haploid chromosome number estimated to be in the region of 220. With regard to chromosome numbers amongst animals, two matters are of particular interest.

First, those animals with very high chromosome numbers usually have small **genomes**, that is the total amount of DNA embodied in a haploid set of their chromosomes is small. Therefore, the individual chromosomes of such animals are very small indeed. Hence the uncertainty about exact chromosome numbers in species like *Lysandra atlantica;* many of its chromosomes are almost too small to see. Beyond that there is no relationship between genome size and chromosome number; indeed, as we shall see later, the animals with the largest genomes have relatively few chromosomes. The animal that probably has the largest genome of all, a small salamander that lives in the tropical rain forests of Central America, has a haploid chromosome number of just 13.

Second, within a particular taxonomic group of animals there may be quite a wide range of chromosome numbers from one species to another. Men and chimpanzees have different chromosome numbers, different species of *Drosophila* have different chromosome numbers and one species of the African clawed toad, *Xenopus laevis,* has 18 chromosomes while another species of the same genus has 108. What can we learn from these observations? Not much at this stage. For the moment I wish only to emphasize that the animal's haploid chromosome number is a fundamental property of that organism's genetic makeup. It represents the minimum number of blocks of genes that remain linked together and segregate together at meiosis. It is a character, a phenotypic character, and in precisely the same sense as is body size, shape and colour, and like all other morphological traits it is a product of evolution through natural selection. Therefore, we should take a special interest in the manner in which natural selection operates to produce a **karyotype**, a chromosome set that contains a particular amount of DNA, and a particular distribution of specific gene sequences that regularly recombine according to a broadly predetermined pattern.

CHROMOSOME SHAPE

At the end of G2 the chromosomes condense and they become visible as objects whose shapes and sizes are characteristic of the species. Generally speaking, a

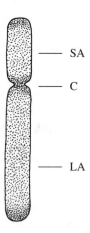

Figure 1.1 A submetacentric chromosome illustrating the primary constriction at the site of the centromere (C) and the long (LA) and short (SA) arms of the chromosome.

chromosome in mitotic metaphase has only two distinguishing features, its length and a transverse constriction that marks the position of the **centromere**.

The portions of the chromosome on either side of the centromere constriction are referred to as the **chromosome arms** (Figure 1.1). In certain circumstances it is possible to introduce other distinguishing features by treating the chromosomes in such a way that they become reproducibly cross-banded in appearance, but that need not concern us for the moment. From length and centromere position we can calculate three factors, the **centromere index**, the **arm ratio** and the **relative length** of the chromosome. The first two of these, centromere index and arm ratio, tell us about the chromosome itself. The relative length tells us about the size of the chromosome in relation to others in the same chromosome set. Centromere index is defined as:

$$\frac{\text{The length of the shorter of the two chromosome arms}}{\text{The length of the whole chromosome}} \times 100$$

It is therefore expressed as a percentage, essentially the percentage of the whole chromosome that is represented by its short arm. Arm ratio is defined as:

$$\frac{\text{Length of the long arm}}{\text{Length of the short one}}$$

It is therefore always greater than 1 and is an alternative method to centromere index for describing the position of the centromere and the relative lengths of the two arms. The relative length is:

$$\frac{\text{Length of the whole chromosome}}{\text{Total length of all chromosomes in the haploid set (including the one being measured)}} \times 100$$

Once again, it can be expressed as a percentage and it gives us some indication of the proportion of the whole genome of the organism that is represented by that particular chromosome. However, in this respect it must be remembered that it is a linear measurement, not a measure of mass or volume.

According to an internationally agreed convention, chromosomes with different centromere positions are given different names. For example, a chromosome with its centromere right in the middle and having two arms of

Table 1.1 Chromosome classification by centromere position

Centromere position	Alternative terminology	Chromosome symbol	Centromere index range
Nearly median	Metacentric	m	46–49
Submedian	Submetacentric	sm	26–45
Subterminal	Acrocentric	st	15–30
Terminal	Telocentric	t	< 15

identical length is called **metacentric**. One with its centromere at its extreme end is called **telocentric**. The agreed terminology is listed in Table 1.1.

At metaphase the chromosomes often, indeed usually, appear double. A metacentric or a submetacentric chromosome will look like a little X (Figure 1.2a, b), with the two chromatids held together at their centromeres; a telocentric chromosome will look like a little upside-down V (Figure 1.2c). What is achieved subsequently is the separation of the **chromatids** that make up these double chromosomes and their orderly distribution as daughter chromosomes to two opposite regions of the cell, a process that involves a structure commonly referred to as the **mitotic spindle**, and a process that involves intracellular movement. Think for a moment about this movement, and think about chromosome size.

CHROMOSOME SIZE

Chromosomes come in very different sizes. Examples of what are probably close to the two extremes of chromosome size amongst animals and plants are shown in Figure 1.3. In numerical terms, the volume of the largest chromosome in the crested newt is roughly 4000 times that of the fourth chromosome of the fruit fly *Drosophila melanogaster*. What does this mean in terms of the physics of chromosome movement? What is the speed of chromosome movement? Do large chromosomes move more slowly than small ones on the mitotic spindle? What are the sizes of the cells through which the chromosomes move? Do big chromosomes need correspondingly big cells?

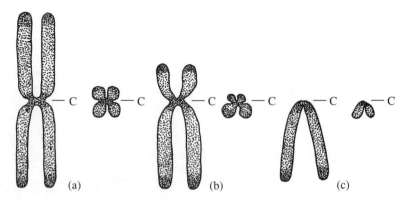

Figure 1.2 (a) The appearance of a large and a small metacentric chromosome at mitotic metaphase with the two chromatids joined at the centromere (C) to form an X-like configuration. (b) The same for a large and small submetacentric chromosome. (c) The same for a large and small telocentric chromosome, forming an inverted V-like configuration.

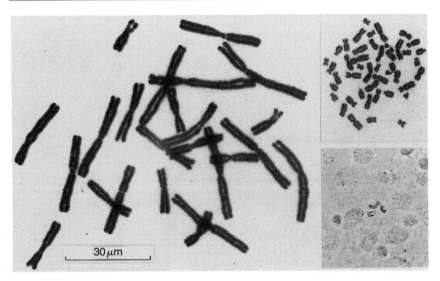

30 μm

Figure 1.3 Chromosome sets from the crested newt (*Triturus cristatus*) (left), man (top right) and *Drosophila melanogaster* (bottom right), each photographed and reproduced at the same magnification. The two tiny fourth chromosomes of *Drosophila* can be seen as a pair of dots just to the left of the centre in the space between the three pairs of larger chromosomes. (Reproduced with the permission of The Royal Society from Macgregor, H.C., *Phil. Trans. Roy. Soc. Lond. B*, **283**, 309–318, 1978.)

CHROMOSOMES IN MITOSIS

In so far as chromosomes are components of the mitotic apparatus, what do they do? What do they contribute to the events of metaphase and anaphase? Are they, as Dan Mazia, one of the pioneers of mitotic science, described them, like a corpse at a funeral, the centre of the proceedings but playing no part. The chromosomes are attached to the mitotic spindle at a specific point, the centromere. What is a centromere? Has it a structure? Has it any characteristics that lead us to suppose that it actually makes spindle material? Is it an active region of the chromosome or is it just the handle of the coffin? A great deal is now known about centromeres, and towards the end of this book I shall introduce some of the latest and most exciting developments in this field of investigation.

At metaphase the chromosomes are attached to a spindle, generally in a ring around the equator of the spindle. Why should they be arranged in a ring (Figure 1.4)? What of the arrangement of chromosomes with respect to one another in this ring? Is it random? Why is there generally nothing in the middle of the ring? The ring lies midway between the two poles of the spindle. Why midway? Why equidistant from each pole?

At anaphase, the chromosomes move apart. One chromatid goes to each pole of the spindle. There are lots of questions that we can ask about this process.

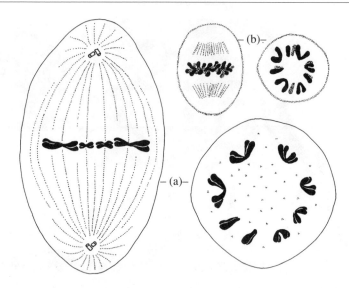

Figure 1.4 (a) Side and polar views of a mitotic metaphase spindle carrying eight chromosomes, each with two chromatids, arranged around the equator of the spindle and half-way between the poles. (b) Artist's impression of the same arrangement as seen with a light microscope in a histological section or an incompletely squashed preparation where chromosomes lie in different focal planes, the centrioles are not visible and individual bundles of spindle fibres cannot be resolved.

Do the chromosomes move in straight lines? Do the chromosomes move away from one another as if they are generating their own motile force, or do they move towards the spindle poles as if they are being drawn apart by forces generated at the poles? Do the chromosomes move with a random orientation, or do they regularly move with their centromeres in the lead? Do sister chromatids always end up at different poles and, if they do, then what determines that one of a pair of chromatids goes one way and the other goes the other way? Do all the chromosomes move at the same speed? The answers to most of these questions are known, but my intention at this stage is not simply to provide answers but rather to draw attention to the importance of the questions themselves.

We could go on and on asking questions about mitosis and the mechanics of cell division. The significance of these questions in the present context is simply that chromosome division by means of the mitotic apparatus is a precise and finely balanced process. It can and does go wrong from time to time, but considering its frequency it is remarkably failsafe. When mitosis combines with certain major structural changes in the chromosomes we get interesting and sometimes heritable changes in karyotype, so that mitosis can be regarded as a major instrument in chromosome evolution. This is perhaps the main reason that we pay attention to the details of its mechanics.

DNA PER NUCLEUS (C-VALUE)

Now so far I have introduced a number of questions regarding the events in cell division, and I have dealt with three major characteristics of chromosomes: their number, their shapes and their sizes. There is one more parameter that should be introduced at this stage. It is the amount of DNA in grams that is present in one complete haploid set of an organism's chromosomes, commonly referred to as the **C-value** or genome size.

It is now more than 40 years since the first studies in which the amount of DNA in individual cell nuclei was measured by determining the amount of DNA in a suspension containing a known number of isolated nuclei and calculating the average DNA content of a single nucleus. The hypothesis was then put forward that DNA, a permanent component of the nucleus and the 'chief component of the hereditary material' in the chromosomes, must be present in the same amount in all the cells of an animal. This hypothesis was, of course, substantially correct, but, as we shall see later, it needed some careful qualification in the light of new evidence on chromosome organization and behaviour at the molecular level.

Studies of C-values have proceeded along two main lines. There have been careful and detailed biochemical measurements of DNA per nucleus employing common laboratory animals such as rats, mice, frogs and fruit flies. These measurements have, to a large extent, provided reference standards that allow comparative data obtained from other species to be converted to absolute amounts. But the biochemical approach to the absolute measurement of DNA per nucleus is a lengthy and expensive one, and it is only well suited to situations where it is possible to obtain and count cell nuclei that are more or less uniform in that they all have just two sets of chromosomes. For example, it is relatively easy to measure the DNA per diploid nucleus in a frog or a chicken by direct biochemical means because both these animals have red blood cells with nuclei in them; so we take a sample of blood, burst the cells, collect and wash the nuclei, count them and measure their DNA. It is much harder to determine the C-value of a mosquito, a sea urchin or even a mammal, for uniform populations of cells that yield good suspensions of clean, countable diploid nuclei are quite hard to find.

The biochemical approach to DNA measurement was quickly followed by a cytochemical one in which it was possible to measure the DNA content of individual cell nuclei using a microscope and a preparation of whole cells on a microscope slide. The approach exploited the Feulgen reaction, which had been introduced into cytology in 1924. The reaction, correctly applied, is absolutely specific for DNA. The reagent, known as **Schiff's reagent**, is formed by the reaction of sulphur dioxide with the dye basic fuchsin to produce a colourless compound known as 'leuco-basic fuchsin'. If this latter substance is then reacted with aldehydes, then a new coloured substance is generated. In histochemical practice, the necessary aldehyde groups are produced by mild hydrolysis of the deoxyribose sugar component of DNA with dilute hydrochloric acid (Figure 1.5). The same effect is not produced by RNA, besides which the acid hydrolysis step that is needed to make the DNA reactive with Schiff's reagent is sufficient to remove most RNA from cells and tissues.

Schiff's reagent, otherwise known as Feulgen dye, is bound strictly quantitatively to DNA, a fact that is best demonstrated by actually trying the technique out on known situations. For example, diploid nuclei bind twice as much Feulgen dye – in practice they have twice as much pinkness in them – as haploid nuclei, tetraploids twice as much again, octaploids twice as much again, and so on. No histochemical test commands the same level of confidence from the quantitative standpoint.

The measurement of Feulgen dye bound to nuclei and chromosomes is generally carried out with a microscope photometer, employing light of around 560 nm wavelength, a green band that is maximally absorbed by the reddish pink colour of the Feulgen dye. A fuller account of the application of Feulgen photometry and other methods for measuring DNA can be found in Chapter 10 of Macgregor and Varley (1988).

The Feulgen reaction and related technologies have been applied to hundreds of species with the single objective of measuring C-values. The standard approach is to make a preparation of cells from one organism whose C-value is unknown in one place on a microscope slide, and another similar preparation from an animal whose C-value is well known nearby on the same microscope slide. The two preparations are then processed together in precisely the same manner. The Feulgen dye bound to nuclei in both preparations is measured, and the unknown C-value is calculated with reference to the known one.

Cells that are commonly used as reference standards when measuring C-values are the red blood cells (erythrocytes) of certain species of frog, newt or salamander. These cells are readily obtainable without killing an animal, they are highly uniform and easy to prepare for the Feulgen reaction and, of course, they all have cell nuclei, unlike the erythrocytes of mammals which are non-nucleated. *Xenopus laevis*, the African clawed toad, has the best established C-value amongst the Amphibia (6×10^{-12} g diploid). Other species that can be used as standards are man, the fruit fly *Drosophila melanogaster*, the common laboratory mouse or rat, and a few species of tailed amphibians.

The accumulated data on C-values are now considerable, although as with chromosome numbers they are meagre in comparison with the total number of known animal species. About 500 species of vertebrate have been measured, and roughly half that number of invertebrates. The species that have been measured are probably broadly representative, although that may not be entirely fair in so far as they represent only about 0.05% of all known species. Moreover, as we shall see later, the word 'representative' may be somewhat misleading, since even closely related species can have widely different C-values.

Why bother about C-values? I offer three answers to that question. First, it is absolutely essential to know the C-value (**genome size**) of an organism before

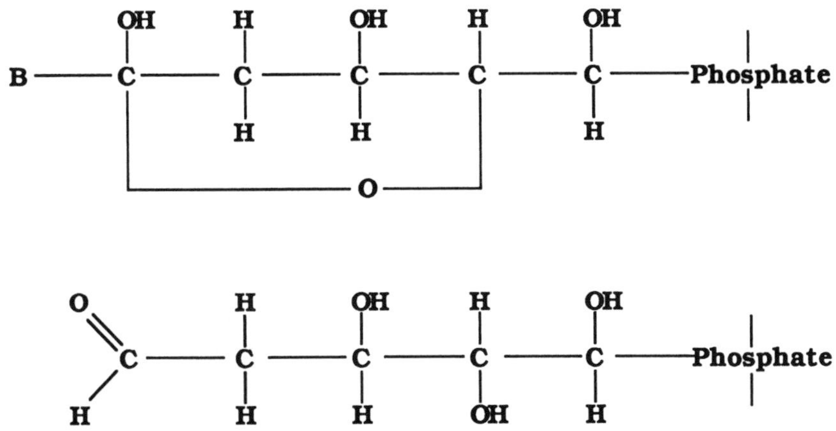

Figure 1.5 The chemistry of the Feulgen reaction, showing a deoxyribose sugar linked on the right to a phosphate and on the left to a purine or pyrimidine (B). Hydrolysis with hydrochloric acid detaches the purine or pyrimidine, leaving a tightly bound aldehyde group (left end of lower diagram) to which leuco-basic fuchsin binds and converts to a pink-coloured compound.

you embark on any studies of molecular genetics. Genome size allows us to decide on probable distances between genes on a chromosome, numbers of copies of genes that are repeated, the actual experimental conditions that must be used to dissect a genome and the probability of finding and isolating a particular gene – essentially the 'needle in the haystack principle'. Knowledge of genome size allows us to assess what we are up against when it comes to taking a genome apart.

Secondly, knowledge of genome size can give us valuable clues to phylogeny and it can expose all kinds of interesting aspects of genome evolution. If two related species have genome sizes *and* chromosome numbers in the ratio of 2:1, we immediately suspect that these species have diverged through one of them doubling its chromosome number to become **polyploid**. In some cases two closely related species may have widely different genome sizes but the same chromosome numbers, which leads us to ask questions about how genomes expand or contract, how chromosomes grow or shrink and how DNA sequences are added or taken away from chromosomes in the course of evolution.

Thirdly, since genome size is rather well related to cell volume (the bigger the genome the bigger the cells), it is clear that genome size must have considerable developmental significance. The ancient Egyptians built their monuments with blocks of stone the size of the modern motor car. Today we build with 8×4 inch bricks. The building techniques are correspondingly different and the buildings themselves reflect the nature of the pieces used to construct them. The fruit fly *Drosophila melanogaster* and the crested newt *Triturus cristatus* provide a close analogy, with genome sizes of 0.15 pg and 20 pg respectively. What is more, genome size is related to **cell cycle time**: the larger the genome the greater the time between cell divisions (mitoses).

RECOMMENDED FURTHER READING

Gorbsky, G.J. (1992) Chromosome motion in mitosis. *BioEssays* **14,** 73–80.

Macgregor, H.C. and Varley, J.M. (1988) *Working with Animal Chromosomes*, 2nd edn. John Wiley & Sons, Chichester and New York.

Pienta, K.J., Getzenberg, R.H. and Coffey, D.S. (1991) Cell structure and DNA organization. *Critical Reviews in Eukaryotic Gene Expression* **1**, 355–385.

Karyology and evolution | 2

2

HISTORICAL PERSPECTIVE

Karyology, or the study of chromosome sets, did not, of course, begin with the study of human chromosomes, but it has certainly been accelerated by our eagerness to see our own chromosomes and map our own genome. It therefore seems appropriate to approach the subject historically, tracing the developments in human cytogenetic technique over the past 40 years and see how these techniques have been applied to the study of chromosomes from other animals with problems in mind that are more fundamentally scientific than medical in nature.

Prior to 1950, most studies of human chromosomes had been carried out on sectioned material. Testis was recognized as a tissue that included a lot of dividing cells, so pieces of testis from dead persons were preserved in a histological fixative and then embedded in wax and sliced into sections thin enough to mount on microscope slides and examine with a normal light microscope. Cells in division could be recognized, and, bearing in mind that an entire set of chromosomes on a mitotic spindle might not all be included in one section, it was possible to obtain a reasonably accurate impression of the total number of chromosomes in the set. The human haploid chromosome number was said to be 24, which happened to agree with the number found in our nearest relatives, the orang-utan, the chimpanzee and the gorilla.

The first real step forward from this position came in 1952, and it was remarkable in three respects. First, it employed skin and spleen from a 4-month-old male fetus, and in so far as good chromosome spreads were obtained from skin cultures it seemed likely that it would soon be possible to study chromosomes from mature live adults. Secondly, the chief investigator, T.C. Hsu, obtained some excellent chromosome spreads in which every chromosome was distinct and countable. How did he do it? The paper that he published in 1952 has the following addendum:

It was found after this article had been sent to press that the well spread metaphases and mitotic anaphases were the result of an accident. Instead of being washed in isotonic saline, the cultures had been washed in hypotonic Tyrode solution before fixation. Furthermore, it was found that Dr. Arthur Hughes of the Strangeways Research Laboratory in England had been carrying out experiments on the effect of hypotonicity upon dividing cells, and his findings were almost identical with ours. We owe our sincere thanks to Dr. Hughes for allowing us to read his original manuscript prior to its publication.

So, one of the most important advances in mammalian cytogenetic technique was the result of an accident. Thirdly, although Hsu was able to examine human chromosomes in complete and well-spread preparations, he nevertheless reported a diploid chromosome number of 48!

Hsu's contribution to early cytogenetic technique was in showing that treatment of cells with hypotonic saline before fixation causes them to swell. The swelling pulls the chromosomes apart so that when they are later placed on a microscope slide they generally lie apart from one another and are individually distinguishable – an important consideration when you wish to pair off and characterize each of 46 small and relatively featureless objects.

The next advance came in 1956, when two Scandinavians, Tjio and Levan, applied Hsu's technique to connective tissue cells from lungs of human embryos cultured in bovine amniotic fluid. The cultures all came from legally aborted fetuses, 10–25 cm in length.These authors added the drug colchicine to their cultures about 12 hours before making the chromosome preparations, and since colchicine stops cells in metaphase of mitosis, they were able to enrich their cultures in cells of this kind. Tjio and Levan then used colchicine on human cells grown in culture and employed Hsu's hypotonic technique to get good chromosome spreads. The conclusion to their paper is particularly interesting and significant, and it represents one of the most extraordinary pieces of scientific diplomacy that has ever been published. Essentially, and most apologetically, they suggest for the first time ever that the long-established chromosome number of 48 for man might be incorrect. They end their paper by saying:

> We do not wish to generalise our present findings into a statement that the chromosome number of man is 2n=46, but it is hard to avoid the conclusion that this would be the most natural explanation of our observations.

Two further developments came in quick succession. Ford and Harnden in England obtained cells from human bone marrow and from skin biopsies and made chromosome preparations using techniques that had been introduced by earlier investigators. In this way they established that it was possible to obtain preparations from live adult males and females. Then Hungerford, working in the United States, demonstrated that human white blood cells could be stimulated to come into mitosis by a haemagglutinating drug extracted from kidney beans – **phytohaemagglutinin**. This opened the way for a full-scale and

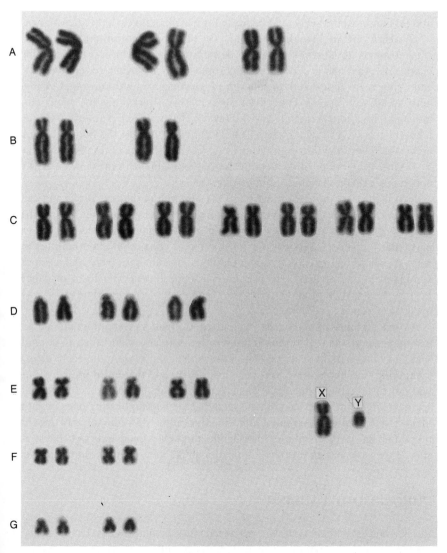

Figure 2.1 Normal male diploid chromosome set with chromosomes arranged in groups, according to relative length and centromere position, and in order of decreasing length within each group, except for chromosomes X and Y. Note that group B chromosomes are submetacentric and groups D and G are acrocentric.

critical analysis of the human chromosome set. Live cells could be obtained from live people, by simply and painlessly removing a small blood sample. They could be cultured, the lymphocytes could be stimulated to divide, the culture could be enriched in dividing cells, and high-quality preparations of complete mitotic chromosome sets could be made for examination by light microscopy.

The early 1960s saw the accumulation of enough information on human chromosomes to allow conclusive demonstration of chromosome number, characterization and grouping of the chromosomes according to relative length and centromere index and correlation of certain gross chromosomal changes with some well-known congenital abnormalities, including the main chromosomal situations associated with Down's (trisomy 21), Turner's (XO) and Klinefelter's (XXY) syndromes. The 23 chromosomes of the human complement were classified into eight groups (A to G and the sex chromosomes X and Y) (Figure 2.1), but positive identification of individual chromosomes was, with only a few exceptions, almost impossible.

Two factors set the trends and goals for research in the late 1960s and early 1970s. First there was the problem of sorting out the chromosomes beyond their groups. The large C-group, with seven chromosomes assigned to it, was particularly difficult. Second, most of the major congenital abnormalities that were known to be linked with major chromosomal changes were of deep social significance, and at least one was known to be linked to a form of cancer. Positive recognition of all chromosomes, and even of chromosome arms, was a must. How was it to be done?

At first, all kinds of approaches were tried. Computers were programmed with the shapes and sizes of normal chromosome sets and then asked to examine scans of other sets in search of aberrations. But those were the early days of computing, the hardware was formidably costly and the entire project simply was not sustainable. Advantage was taken of the fact that certain chromosomal segments replicate later than others, such that the late-replicating ones might be identified by autoradiography after incorporation of radioactively labelled thymidine into cells in culture. In principle, this was a good approach, but it was slow, expensive and unrewarding in relation to the investment of time and effort.

CHROMOSOME BANDING

In 1968 and 1970 there were two major technical breakthroughs. On the basis of his long experience with microdensitometry of cells and tissues, Caspersson – who indeed might be regarded as the founder of all microdensitometric technologies in biology – reasoned that it was possible to make accurate and reproducible microphotometric measurements on objects down to 0.3 µm in diameter. He then pointed out that **ultrafluorometric** techniques, employing fluorescing dyes such as acridine orange, were good enough to measure the spectral emissions of acridine bound to droplets of DNA containing as little as 10^{-15} to 10^{-16} g of DNA. A small human chromosome contains about 10^{-13} g of DNA, so that microfluorimetry should certainly allow the detection of one-tenth of that amount. Caspersson went on to argue that an average gene might be expected to contain between 400 and 4000 nucleotide pairs. At 650 daltons per nucleotide pair, 200 copies of an 'average gene' clustered together at one point on the chromosome would make very nearly 10^{-16} g of DNA. There were reasons in 1968, just as there are now, for supposing that certain gene sequences might be repetitive up to several thousand fold and that in some cases all members of a family of repeats would be clustered together in one place on a chromosome. Caspersson considered certain fluorescing substances known as alkylating agents which attach to the N7 atom of the guanine ring, and he wondered whether certain of these substances might not bind preferentially to G+C (guanine + cytosine)-rich sequences on metaphase chromosomes. Naturally, he also took account of the fact that some alkylating agents are fluorescent dyes! In the end he took quinacrine and quinacrine mustard and used them to stain mitotic chromosomes from humans. The results were astonishing and immediately applicable to the problem of distinguishing all the chromosome arms in the human set. The chromosome-specific cross-banded patterns that follow

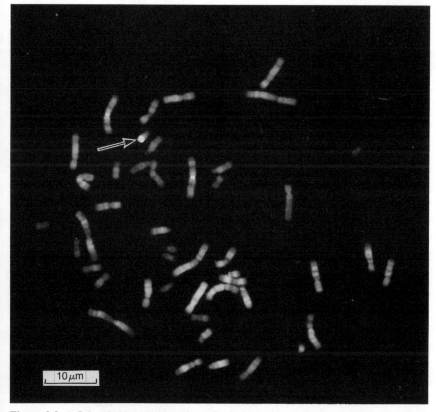

10 μm

Figure 2.2 A Q-banded human metaphase. Each chromosome can be positively identified, and the Y is especially prominent owing to the bright fluorescence of its long arm (arrow). (Photograph kindly provided by Dr M. Schmid of the University of Wurzburg.)

staining with either quinacrine or quinacrine mustard are now referred to as
'Q-banding' (Figure 2.2). Notwithstanding the undoubted value of Q-banding
as a means of identifying human chromosomes, it does have its drawbacks.
Perhaps most importantly, it requires the use of a microscope equipped for
fluorescence microscopy and it is more difficult to photograph than a
conventional image; but it was hailed as a major step forward for human
cytogenetics and it was the essential forerunner of further series of important
developments in the field. Incidentally, Caspersson's original prediction that
quinacrine would allow detection of G+C-rich DNA sequences was fully
vindicated. A 5% shift in the G+C bias in DNA is enough to produce a 50%
change in quinacrine fluorescence.

A short time later, Gall and Pardue were experimenting with the technique of
in situ **nucleic acid hybridization**, and as one of their early experiments they
denatured mouse chromosomes with alkali, making the chromosomal DNA
single stranded rather than double helical, and then incubated them in a 0.3 M
salt solution containing purified highly repetitive A+T (adenine + thymine)-rich
mouse 'satellite' DNA. The important point is that Gall and Pardue stained their

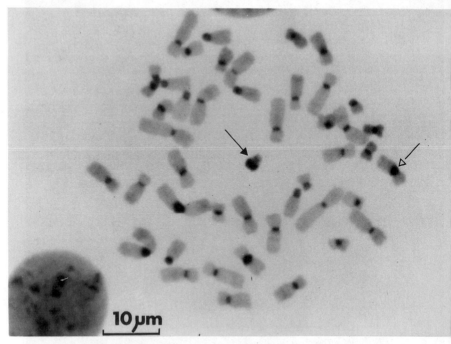

Figure 2.3 A C-banded human metaphase. The centromere regions of all the
chromosomes are darkly stained (open arrow), as is the long arm of the Y chromosome
(solid arrow). (Reproduced with the permission of John Wiley & Sons from Macgregor,
H.C. & Varley, J.M., *Working with Animal Chromosomes*, 2nd edn, 1988.)

preparations with **Giemsa**, and this produced a clear differential staining of the centromere region of each chromosome.

Sumner, Evans and Buckland, working in the Medical Research Council's cytogenetics laboratory in Edinburgh, proceeded to do the same thing with human chromosomes (denaturing, incubating in saline and then staining with Giemsa), and they obtained the same results. Each centromere was differentially stained and, what is more, each chromosome was distinguishable on account of the amount of densely staining material at its centromere. Here was a relatively quick, simple technique that required no more than a few inexpensive reagents and an ordinary microscope. Centromeric staining has since been designated **C-banding** (Figure 2.3). The technique is widely applicable, indeed more so than any other chromosome banding technique, and it is easy to carry out and reproduce.

In the course of their investigations, Evans and his colleagues noticed a faint transverse banding of chromosome arms in some of their preparations, and they

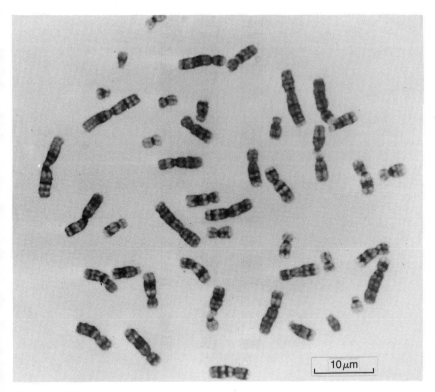

Figure 2.4 G-banded human metaphase. The chromosomes are differentiated into a pattern of light and dark bands, allowing unequivocal identification of each chromosome pair. (Reproduced with the permission of John Wiley & Sons from Macgregor, H.C. & Varley, J.M., *Working with Animal Chromosomes*, 2nd edn, 1988.)

promptly set about developing a method that would enhance this effect. The outcome was the 'ASG' technique, the letters representing the words Acetic, Saline and Giemsa. The chromosomes were first fixed with a mixture of methanol and acetic acid. They were then incubated at 60°C for 1 hour in a mixture containing 0.3 M sodium chloride and 0.03 M sodium citrate (commonly known as 2 × SSC, the letters in this case representing standard saline citrate), and they were finally stained with Giemsa. After staining with ASG it is usual for human chromosomes to appear extensively cross-banded, with each chromosome arm having its own distinctive and entirely reproducible pattern of **G-banding**.

G-banding is rightly accepted as being one of the most useful and reliable techniques in mammalian karyology. It allows the dependable identification of all chromosome arms, and even of specific and sometimes quite short segments of arms. It is now more than 20 years since its invention and there have been many variations and improvements in the technique (see Recommended further

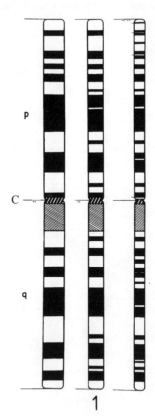

Figure 2.5 The G-banding patterns on human chromosome 1 at three different levels of resolution. The left chromosome shows the level that reveals 400 bands per complete haploid chromosome set. The middle one is the 500-band level and the right-hand one the 850-band level. C, centromere; p, short arm; q, long arm.

reading). The human G-band pattern has been analysed in the greatest possible detail and is subject to a complex set of strict internationally agreed nomenclatural conventions (see Chapter 11).

Other methods of producing chromosome banding have been introduced. **High-resolution G-banding**, obtained by deliberately working with prophase or early metaphase chromosomes that are not yet fully contracted, produces over 800 bands in the human karyotype (Figure 2.5).

R-banding, a technique that exploits the constant and asynchronous pattern of replication of chromosomes, has proved especially valuable in allowing high-resolution chromosome identification in species whose chromosomes will not, for one reason or another, respond to G-banding. *Xenopus laevis*, for example, the most popular of all research amphibians, has chromosomes that have not responded at all well to attempts at G-banding, yet high-resolution R-banding produces over 650 bands, permitting the most detailed karyotypic analyses and opening the door in this important genus for in-depth studies of chromosome organization and evolution.

What can be done with modern techniques of chromosome identification? Chromosome arms can be identified unequivocally. Structural rearrangements (translocations, inversions, deletions, duplications), even quite small ones, can be identified with ease. Karyotypes can be compared at a level of detail that is, of course, quite impossible with uniformly stained chromosomes. Perhaps the greatest limitation of G-banding is that it only seems to work on mammalian chromosomes. Chromosomes from other groups of organism are either not differentiated in a manner that will permit G-banding or cannot be prepared in ways that preserve the complex that stains differentially with Giemsa. Nevertheless, there are now many different technologies labelled 'chromosome banding' and there are very few instances in which it is impossible to produce good linear differentiation of metaphase chromosomes such as will permit detailed descriptions of karyotypes, detection of structural rearrangements and useful comparisons between species. It is probably fair to say that techniques of chromosome banding have gone about as far as they can go and there is probably not much more that we can or need to do to develop them further. The latest trends are towards cytological location of individual genes and specific DNA sequences, employing elegant and rewarding techniques of molecular biology.

KARYOTYPES, GENOMES AND EVOLUTION

The reasons for studying the chromosomes of man are largely medical. We cherish the hope that by understanding better the chromosomal abnormalities that are associated with a variety of distressing organic defects we shall perhaps be better equipped to deal with the medical and sociological problems surrounding these defects. Very impressive progress has been made, and the human-genome sequencing programme coupled with the latest techniques of 'gene therapy', is likely to form a basis for some interesting, albeit highly expensive, approaches to the management of a range of certain serious and hitherto incurable illnesses. The reasons for studying the chromosomes of animals are not at all medical. We look at chromosomes because we wish to know how they are constructed and how they function, and they are exceedingly useful as relatively simple features, closely representing the organism's genotype, that can give valuable clues to phylogeny, evolution and taxonomic relationships.

The methods that are employed in the study of chromosomes use three main characteristics: size, shape and linear differentiation. So far we have considered size in terms of DNA content and chromosome length, we have considered shape in terms of centromere index and arm ratio and we have considered linear differentiation in terms of quinacrine and Giemsa banding. Later we shall be looking at linear differentiation in the structural sense when we come to consider giant polytene and lampbrush chromosomes. In the meantime let us confine our attention to chromosomes in mitosis and see what we can do with them.

With respect to size alone there are some extremely odd cases: animals that belong to the same genus, that have the same chromosome numbers and the same karyotypes (virtually identical chromosomes with respect to centromere indices and relative lengths), but have widely different C-values, signifying that corresponding chromosomes in one species are much larger than their counterparts in a related species. *Gammarus pulex*, for example, a small freshwater amphipod crustacean, has 26 or 27 chromosomes containing three times as much DNA as the 26 chromosomes of its near relative *G. chevreuxi*. *Mesostoma ehrenbergi*, a small flatworm (Turbellaria, Rhabdocoela), has four chromosomes with 11 times as much DNA as the four chromosomes of its relative *M. lingua*. Among the plants we have genera like *Allium*, belonging to the onion or garlic family, in which there are 13 species with a chromosome number of 16 and a range in C-values from 15 to 33 pg of DNA. What is even more remarkable is that species of *Allium* that have widely different C-values and hence chromosome sizes can form fertile hybrids, which means that meiosis must involve synapsis and exchange between chromosomes of different lengths.

Two more examples deserve special mention. First there are the plethodontid salamanders, and particularly those belonging to the genus *Plethodon*. Here we can see three different species that are known to have been separated for around 50 million years, that have virtually identical karyotypes, and yet have C-values in the ratio of 1:2:3. In other words, we are looking at a situation in which each member of a set of 14 chromosomes has, in the course of millions of years of evolution, retained exactly the same shape and relative size, but each has increased (or decreased) in absolute size by two or threefold. What might this tell us about the adaptive value of that particular karyotype or about the molecular mechanisms that lead to changes in the sizes of chromosomes by addition or removal of DNA sequences?

Lastly in this context, there is the case of *Chironomus*, where two subspecies may be different in their C-values by as much as 30% and yet have identical chromosomes except only for certain chromosome loci that have 2, 4, 8 or 16 times as much DNA in one species than in the other.

The general message that comes over from these examples is that evolution can generate quite striking changes in the sizes of chromosomes without affecting their relative dimensions within a chromosome set, and without greatly affecting the morphology or nature of the organism. How do such changes come about and how can they be tolerated?

When we come to consider chromosome shape we are, of course, concerned with differences in the relative dimensions and centromere positions, not necessarily accompanied by any changes in chromosome size or C-value. Two acrocentrics fuse in the neighbourhood of their centromeres to form a single metacentric, or a metacentric is cleaved at its centromere to form two acrocentrics (events of this kind involving rearrangements of whole chromosome arms are commonly known as 'Robertsonian' changes), or an acrocentric undergoes an inversion that transposes its centromere from the end to the middle of the chromosome. These kind of events have been commonplace amongst animals, and in some cases, notably *Drosophila*, whole phylogenies have been reconstructed on the basis of putative sequence of karyotype reshuffling.

The study of chromosome form and species kinship becomes closely linked with the study of linear differentiation just as soon as we come to consider chromosomes that can be Q-, G- or C-banded. One good example of a karyotypic similarity between kindred species concerns goats, sheep and cattle. In *Capra hircus* (goat) the diploid chromosome number (**2N**) is 60 and the chromosome set comprises 29 pairs of **autosomes** and two **sex chromosomes** (Figure 2.6). All of these chromosomes are acrocentrics. *Ovis aries* (sheep) has a diploid number of 54, comprising 26 pairs of autosomes and two sex chromosomes. Of these 26 autosomes, 23 are acrocentrics and the three largest chromosomes are metacentric (Figure 2.6). In *Bos taurus* (ox or cow) 2N = 60 and, once again, all chromosomes are acrocentrics. So it would appear that we have three broadly related species and three broadly similar karyotypes with a suggestion that there have been some Robertsonian fusions in sheep to generate the three metacentric chromosomes. The whole situation comes much more clearly into focus when we stain with Giemsa to produce G-banding. That shows us that each of the 23 acrocentric chromosomes of the sheep is represented by an identical chromosome in the goat, and the arms of the three pairs of sheep metacentrics are identical matches to the six remaining goat acrocentrics. The largest metacentric is made up of arms that resemble chromosomes 1 and 3 of the goat, the next is made up of chromosomes 2 and 8 and the third is chromosomes 4 and 9. A similar homology is evident in all but two of the acrocentrics in the ox that have no counterparts in either sheep or goat.

Extensive banding information is now available for the chromosomes of all hominoids, owing mainly to the widespread interest in the phylogeny of man

GOAT

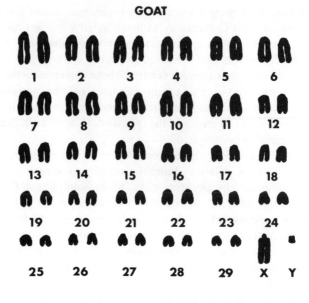

SHEEP

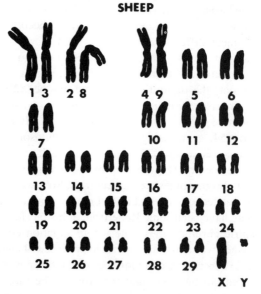

Figure 2.6 Karyotypes of goat and sheep showing the relationship through Robertsonian changes involving telocentric goat-derived chromosomes 1+3, 2+8 and 4+9 to make sheep metacentric chromosomes 1, 2 and 3.

and other species of primate. Man (*Homo sapiens*), the chimpanzee (*Pan troglodytes*), the gorilla (*Gorilla gorilla*) and the orang-utan (*Pongo pygmaeus*) have attracted the most attention. The similarities of the karyotypes of the three 'great apes' and man are quite striking and serve as a basis for some entirely reasonable speculations regarding the phylogeny of the group as a whole and the manner and order in which karyotypic changes have happened in the course of evolution. For example, the X chromosomes of all four species are virtually identical with regard to Giemsa G-band pattern. The same can be said of chromosomes 21 and 22 in man as compared with chromosomes 22 and 23 respectively in the three apes. Chromosome 6 in man is remarkably similar to chromosome 5 in the apes. Chromosome 5 in man resembles chromosome 4 in the apes. Chromosome 1 is similar in all four species apart from a minor shift in banding pattern near the centromere of the human chromosome 1. Indeed chromosome 1 is of some special interest, since it has clear counterparts in the more distantly related baboon (*Papio papio*) and green monkey (*Cercopithecus aethiops*).

But what about the overall difference in chromosome number between man (2N=46) and his closest ape-like relatives (2N=48). The reason could not be simpler: the 12th and 13th longest chromosomes of the chimpanzee and the 11th and 12th longest chromosomes of the gorilla and orang-utan are subacrocentrics whose Giemsa G-band patterns can be matched almost exactly to those of the two arms of the large metacentric chromosome 2 in man (Figure 2.7), another case of a simple Robertsonian change that most likely involved a centric fusion of the counterparts of ape chromosomes 12 and 13 in an ancestral human line, or, less likely, the splitting of a large metacentric to give two smaller acrocentrics in the common ancestor of the three genera of great apes.

With regard to individual chromosomes, the homology amongst man and monkeys goes much deeper than Giemsa G-banding, as has been shown by exploiting a range of techniques of gene mapping. One of the earliest approaches to this problem is, I think, particularly elegant and experimentally ingenious. It requires some preliminary biochemical explanation. First, baboon or green monkey cells are grown in tissue cultures and then made to fuse with mouse tissue culture cells that lack the ability to synthesize an enzyme called thymidine kinase. Without that enzyme the cell cannot utilize thymidine that is supplied to it in the culture medium. The fusion of cells is produced by mixing the cells and treating them with a surface-active agent that causes them to fuse with one another whenever they touch. These **hybrid cells** are cultured in a medium containing aminopterin, a substance that prevents the cells from making their own thymidine, so that if they are to survive they must be able to use exogenous thymidine and to do this they must have thymidine kinase. The gene for that enzyme can only be provided by a primate chromosome. The conditions therefore select specifically for hybrid cells. Lone mouse cells die because they do not have any thymidine kinase and so cannot use thymidine from the culture medium and the aminopterin prevents them making thymidine of their own. Lone monkey cells are specifically killed by including a substance called ouabain in the culture medium, which inhibits an important step in energy metabolism only in primate cells but not in mouse ones. So all we are left with is

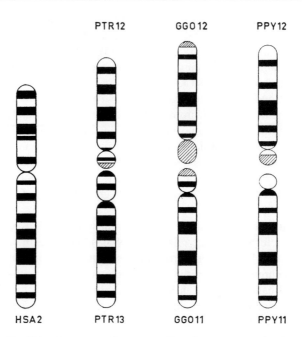

Figure 2.7 The G-band patterns of corresponding chromosome arms from man (HSA), chimpanzee (PTR), gorilla (GGO), and orang-utan PPY. HSA2 is formed by a Robertsonian fusion of chromosomes now represented by numbers 12 and 13 in PTR and numbers 11 and 12 in the other two apes.

mouse–monkey hybrid cells. Cultures of these cells grow and come to possess chromosome sets that contain neither a full mouse karyotype nor a full primate karyotype but a balanced mixture of mouse and primate chromosomes.

A search is then made in these cells for three enzymes that are known to be synthesized from messenger RNAs produced by genes on chromosome 1 in man. These enzymes are characterized biochemically to see if they are the same as those found in man. A search is then made to see if their occurrence always correlates with the presence in the karyotype of the hybrid cell of chromosome arms that have the same Giemsa C-band patterns as those of the two arms of chromosome 1 in man. The correlation is excellent, although not absolute. In green monkey, the banding patterns of chromosomes 4 and 13 add up approximately to those of the long and short arms respectively of chromosome 1 in man, and two of the human chromosome 1 enzymes can be assigned to chromosome 4 and the third to chromosome 13. In baboon, only two of the enzymes can be characterized, for technical reasons, but both can be assigned to the baboon chromosome 1, which is recognizably similar in banding pattern to human chromosome 1. So karyotypically corresponding chromosomes in related species not only have similar Giemsa G-band patterns, they also frequently carry the same genes in approximately the same positions.

So far the emphasis has been on a few clear and interesting cases where kinship is upheld by karyotypic similarity and patterns of Giemsa staining and gene distribution, and of course there are many more examples besides the ones that are mentioned here. However, there are plenty of examples of disagreement and difficulty. Bats, gerbils and muntjac deer are worth mentioning in this regard. The large genus of bats, *Myotis*, shows great karyotypic uniformity, with 18 species having almost identical karyotypes. The bat family Phyllostomatidae, on the other hand, shows enormous karyotypic variation, even between quite closely related species. Chromosome numbers vary from 16 to 46. The number

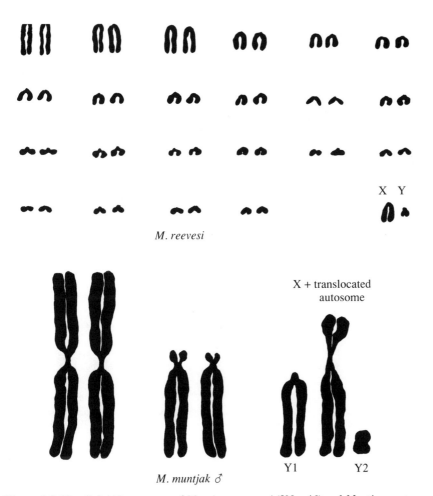

Figure 2.8 The diploid karyotypes of *Muntiacus reevesi* (2N = 46) and *Muntiacus muntjak* (2N = 7 male or 6 female). In *M. muntjak* an acrocentric autosome was translocated onto the original X chromosome but its homologue remained free, thereby giving rise to two Y chromosomes, the longer of which is actually an autosome. Y2 in this drawing is the true Y.

of chromosome arms (*nombre fondamentale*, NF) varies from 22 to 48. The family includes a range of complex sex chromosome arrangements, and there is evidence of a whole series of chromosome repatternings and structural rearrangements. Estimation of kinship on the basis of karyology within this family would be quite impossible.

Amongst the gerbils, one 'superspecies', *Gerbillus pyramidum*, includes forms (species or subspecies) with chromosome numbers as different as 66, 52 and 40, and male F1 hybrids resulting from crosses between different forms show chromosome behaviours at the meiotic divisions in the testes that suggest a widespread scrambling of the karyotype through a series of Robertsonian changes and structural rearrangements of chromosomes. Once again, given the chromosomes alone, estimation of kinship would be impossible.

Certainly the most extreme case of karyotypic variation within one genus is that of the muntjac deer. One species, Reeve's muntjac (*Muntiacus reevesi*), has 2N=46 (Figure 2.8). Its close relative, the Indian muntjac (*Muntiacus muntjak*), has 2N=7 including sex chromosomes (X,Y1,Y2) in the male and 2N=6 in the female (Figure 2.8). The karyotypes look entirely different but the close kinship of the two species is beyond doubt.

At the end of the day, who is to be judged correct when it comes to deciding upon taxonomic matters, the cytogeneticist or the systematic zoologist? In my view, the former can only point the way, but he cannot alone solve the problems of the evolution of taxa unless, of course, he can develop his skills to the point of seeing how the organization and reorganization of DNA sequences in chromosomes is accomplished and how it influences patterns of organic development.

RECOMMENDED FURTHER READING

Chiarelli, A.B. and Capanna, E. (1973) *Cytotaxonomy and Vertebrate Evolution.* Academic Press, London and New York.

Dutrillaux, B, Rethore, M.O. and Lejeune, J. (1975) Comparison of karyotype of orang-utan (*Pongo pygmaeus*) to those of man, chimpanzee and gorilla. *Ann. Genet.* **18**, 153–161.

Evans, H.J., Buckland, R.A. and Sumner, A.T. (1973) Chromosome homology and heterochromatin in goat, sheep and ox, studied by banding techniques. *Chromosoma (Berl.)* **42**, 383–402.

Hsu, T.C. (1952) Mammalian chromosomes *in vitro*. I. The karyotype of man. *J. Hered.* **43**, 17.

ISCN (1985) An International System for Human Cytogenetic Nomenclature: Birth Defects. Original Article Series **21**, No. 1 The National Foundation 1985.

Verma, R.S. and Babu, A. (1989) *Human Chromosomes, Manual of Basic Techniques.* Pergamon Press, Oxford and New York.

Wurster, D.H. and Benirschke, K. (1970) Indian muntjac, *Muntiacus muntjak:* a deer with a low diploid number. *Science* **168**, 1364.

Meiosis

3

BASIC PRINCIPLES

This chapter begins with a formal description of meiosis and follows on with a discussion of three major problem areas in the meiotic cycle: chromosome pairing, chiasma formation and disjunction. Most people would agree that meiosis is quite a complicated process. We can tell that it is old in evolutionary terms since it is used in a more or less common form by such a wide range of animals, plants and other organisms. It is very much a tried and tested process with an enormous level of minor variation. To understand it fully we must first grasp the fundamental evolutionary value of the process and then we must learn to appreciate the significance of the fine-scale details because it is they that will help us more than anything to understand how natural selection has operated at the molecular and supramolecular levels to produce this highly tuned and very important adaptive sequence.

In the beginning it is probably best to think of meiosis as two very distinct processes both committed to the same purpose. Separate these two processes in the mind and the whole complex sequence becomes much more manageable. The first is **random segregation of chromosomes**. The second is **chiasma formation and crossing over**. The evolutionary advantage of both is that they produce a random mixing or reshuffling of chromosomes and genes from one generation to the next. The word meiosis comes from the Greek *meioo,* to lessen, or *meion*, less, and *osis*, a word for a process or condition, signifying a process that lessens or reduces something – chromosomes reduced from diploid to haploid, as we happen to know. It usually involves two divisions known as the first and second meiotic divisions. Random segregation happens at the first division. All your cells contain one set of 23 chromosomes that you received from your mother and one set of 23 derived from your father. At the first meiotic division these chromosomes do not themselves divide as they would do at a mitosis. Instead, they separate into two groups of 23 in an entirely random fashion such that the chances of one of the two resulting cells containing a pure set of your maternally derived chromosomes and the other an entire set of your paternal chromosomes are $1:2^{23}$, which is highly unlikely, to say the least! We must never forget this process. It alone is responsible for promoting an immense level of genetic variation from generation to generation, a level that, of course, becomes greater as the number of chromosomes increases. It alone, in one simple division, has accomplished a reduction to the haploid chromosome number and a reassortment of the chromosomes; but it is not enough. It represents the 'cutting' of the card deck but not the shuffling of the cards. The latter is accomplished by recombination. Recombination is a rearrangement of the genes that are positioned on the same chromosome and it is accomplished by chiasma formation and crossing over.

Chiasma formation and crossing over impose certain requirements. There must be two homologous chromosomes present, so we must start with a diploid cell. These chromosomes must get together, pair up, for chiasmata and cross-overs to form. Cell division is usually preceded by a replication of chromosomes during interphase, such that each chromosome comes to consist of two chromatids. If we retain this feature before the onset of meiosis, then to get down from diploid to haploid we will have to go from four chromatids to one, which means, inevitably, *two* consecutive divisions without an intervening replication of the chromosomes. The fact that there are four chromatids involved in crossing over is good of course. It means that we keep the option of not crossing over, of retaining genes in the same clusters, linkage groups, as exist in the parent generation, whilst still generating new gene combinations in the chromatids that *are* involved in cross-overs. What must be plain to see is that all the apparently mind-boggling complexities of the meiotic process have evolved in coping with four chromatids resulting from a premeiotic replication and the cytological gymnastics needed to accomplish chiasma formation and crossing over. Cutting the pack is easy; shuffling the cards requires a much more elaborate process!

TIMING AND SEQUENCE

The basic sequence in which recombination happens is as follows. The cell that leads into meiosis has the 2C quantity of DNA embodied in the typical diploid number of chromosomes for the organism concerned. The chromosomes in that cell replicate for the last time during the S-phase that comes immediately before the start of the first meiotic division, so that the cell enters meiosis with two sets (diploid) of double chromosomes (two chromatids each). Chromosome number and nuclear DNA content are reduced from diploid and 4C to haploid and 1C by two consecutive divisions, without any intervening S-phase. The first division separates whole chromosomes to make the daughter cells haploid and 2C (chromosomes still consisting of 2 chromatids). The second division separates chromatids to make the cells haploid and 1C.

Meiosis as it occurs in the common newt (*Triturus vulgaris*) is summarized in Figure 3.1. Why choose the common newt? Simply because it has large and very beautiful chromosomes and the timing of its entire meiotic process was carefully worked out jointly by two of the most distinguished chromosome biologists of the century, H.G. Callan and J.H. Taylor (1968). In Figure 3.1 we see the time occupied by each stage from **spermatogonium** to **spermatid**, we see the amount of DNA per nucleus expressed in multiples of the C-value, the numbers of chromosomes expressed as multiples of N and there are notes on certain cytological features and happenings that normally occur at certain points in the meiotic cycle.

One of the earliest and historically most significant discussions of meiosis and chiasma formation, carried out by F.A. Janssens at the turn of the century (1909), was based on studies of an amphibian known as the Californian slender salamander (*Batrachoseps attenuatus*). The description that follows here is a contemporary outline of meiosis in this and certain other species of North

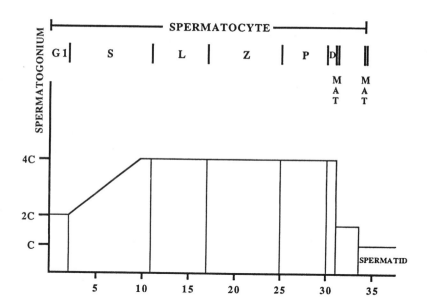

Figure 3.1 The timing of meiosis from the end of the last spermatogonial mitosis through to the end of the second meiotic anaphase in the newt *Triturus vulgaris*. The diagram is presented as a graph of the nuclear DNA content in terms of the C quantity (haploid equivalent) of DNA per nucleus. Spermatogonia begin with the 2C amount of DNA embodied in a diploid set of chromosomes, each of which comprises just one chromatid. Premeiotic S-phase then produces spermatocyte nuclei with the 4C amount of DNA in the diploid number of chromosomes, each chromosome now having two chromatids. The two meiotic divisions then produce a stepwise reduction of the nuclear DNA content to 1C and the chromosome content to haploid, with the formation of spermatids. Note the long first meiotic prophase extending from day 11 through to day 30 or beyond.

American salamander. Most of these animals have haploid sets of 13 or 14 large metacentric or submetacentric chromosomes. Some of them have what might fairly be described as the largest chromosomes in the world! Their testes are long and thin and at certain times of year they show a complete series of spermatogenic stages from spermatogonia at one end to mature sperm at the other. As a source of material for studying the formal cytology of meiosis they are unsurpassed, but they are only realistically available to people in North and Central America, and although most species are incredibly abundant it may only be a matter of time before some begin to appear on endangered species lists.

All the definable stages in the meiotic process are represented in Figures 3.2 to 3.10. Each of these figures presents a photomicrograph of a section through a region of testis where a particular stage is in progress (a), together with a semischematic drawing of some nuclei or groups of chromosomes in that stage (b), followed by a diagram illustrating the main features of the known or supposed form of the chromosomes at that stage (c). All the semischematic drawings in figures b show nuclei from a hypothetical organism with a haploid chromosome number of 4: one large metacentric, one medium-sized sub-metacentric, one small submetacentric and one small metacentric, best seen at the outset in the diploid spermatogonia shown in Figure 3.2b. The photomicrographs (figures a) are all derived from *B. attenuatus* (N=13). The notes to the right of figures c are intended to help with interpretation of the photomicrographs and drawings.

We begin our account of meiosis with a diploid spermatogonium that is one of the cells that result from many rounds of **spermatogonial mitosis** (Figure 3.2a, b and c), the normal mitotic divisions that are responsible for generating the vast numbers of cells needed to sustain the production of millions of spermatozoa in an animal's testis. Accordingly, in Figure 3.2a it is plain to see (although impossible to count accurately) that at metaphase the number of chromosomes on the spindle is certainly nearer to the diploid number of 26 than the haploid number of 13. In Figure 3.2b, eight chromosomes are represented and at metaphase each of these consists of two chromatids. The resulting daughter cells have the full diploid (eight) number of chromosomes. Nothing complicated here: just genuinely normal mitoses, involving the replication, division and equational segregation of the full diploid number of chromosomes.

In the final premeiotic interphase, the spermatogonium passes through a premeiotic S-phase and replicates its chromosomes so that each comes to consist of two chromatids. Then follows the first meiotic prophase, a relatively complicated sequence that has evolved as part of the mechanism that promotes chromosome pairing, chiasma formation and genetic crossing over between chromatids belonging to homologous chromosomes. There are four substages in meiotic prophase, and their names relate to the behaviour and structural characteristics of the chromosomes at each stage.

Leptotene (a word derived from the Greek *leptos*, meaning thin, fine or delicate, and *tainia* a strand) nuclei are shown in Figure 3.3. They are characterized by some large chromocentres that represent the fused centromere regions of all the chromosomes and indistinct, thin and finely granular chromosomes that seem to have no special arrangement within the nucleus, as far as we can tell with the light microscope. Note very particularly that the chromosome strands represented in Figure 3.3c, although they appear to be single, consist of two chromatids that are the products of the last premeiotic S-phase replication.

Zygotene (from the Greek *zugon*, meaning a yoke or joining together and *tainia* a strand) nuclei are shown in Figure 3.4a, b and c. During this stage, homologous chromosomes synapse or join together, and in the particular species illustrated here synapsis commences simultaneously at both ends of the chromosomes and proceeds towards the middle. Throughout zygotene, the ends of all the chromosomes point towards one region of the cytoplasm where the centrioles are located, producing the so-called 'bouquet' arrangement of bivalents that will persist throughout zygotene and the following stage, pachytene. In the zygotene nuclei shown in Figure 3.4 the chromosomes have synapsed over nearly half their lengths, so that one half of the nucleus is occupied by relatively thick strands of the synapsed regions, while the other half is occupied by thinner unsynapsed material. The large fused masses of chromocentric material that we saw at leptotene break up in zygotene to form smaller masses, each of which represents the fused centromeric regions of one **bivalent** (Figure 3.4c).

Pachytene (Greek *pachys*, meaning thick) nuclei are shown in Figure 3.5a, b and c. Synapsis has brought the two members of each homologous pair into such intimate association that the bivalent appears as a single structure in fixed and stained preparations, although its doubleness can sometimes be seen in phase-contrast images of living material. At pachytene, the bouquet arrangement of the chromosomes is particularly clear, the nucleus containing the haploid number of U-shaped loops with their ends still orientated towards the portion of the nuclear envelope that is adjacent to the centriolar complex (Figure 3.5c). Remember that not only do these apparently single pachytene bivalents, as represented in Figure 3.5b, consist of two synapsed homologous chromosomes but each of these chromosomes consists of two chromatids. The whole thing is therefore an intimately associated four-strand structure.

Pachytene is followed by a **diffuse stage** (Figure 3.6a, b and c) during which the nuclear material assumes an exceedingly finely fibrous appearance and almost all trace of the chromosomes' axes is lost. The structure of meiotic chromosomes during this stage is unknown, although it has been suggested that they may have something in common with the lampbrush chromosomes that characterize the post-pachytene stage in growing ovarian eggs (see Chapter 9). However, this analogy should not be taken too far. The spermatocyte nucleus is much smaller than the oocyte nucleus. The chromosomes of diffuse spermatocytes synthesize little, if any, RNA, whereas the lampbrush chromosomes of oocytes are clearly synthetically active. There are distinct axial chromomeres on

Figure 3.2 (a) A section through a small region of a salamander's testis with cells in mid-metaphase of **spermatogonial mitosis**; haploid chromosome number 13. A glance at the group of chromosomes indicated with the arrow is sufficient to establish that it is more likely to consist of 26 rather than 13 chromosomes. This region of testis was at the far anterior end, where the earliest stages in spermatogenesis might be expected to occur. The 25-μm scale bar applies also to Figures 3.3–3.10.

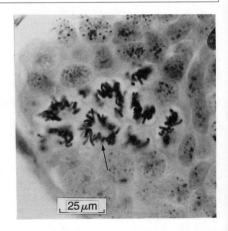

(b) A semischematic drawing of cells in various stages of spermatogonial mitoses in an organism with a haploid chromosome number of 4. *Count the chromosomes.* Each chromosome in the upper left group consists of two chromatids.

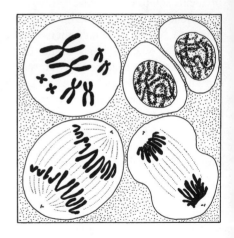

(c) Normal mitotic behaviour showing two chromosomes, each with two chromatids and outward-facing kinetochores, dividing to produce two daughter chromosomes each.

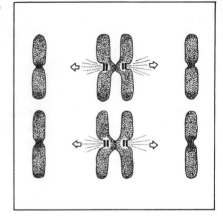

Figure 3.3 (a) A group of **leptotene** nuclei, each showing the disorderly arrangement of finely granular nuclear material and one or a few compact chromocentres (arrows) that represent fused centromeric regions of chromosomes. Like the cells in Figure 3.2a, these leptotene cells were located in the far anterior of the testis where the earliest stages of meiosis occur.

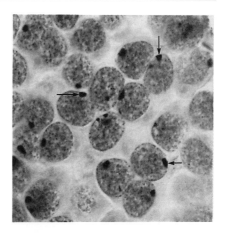

(b) A group of leptotene nuclei illustrating the main features shown in Figure 3.3a. Arrow indicates a chromocentre.

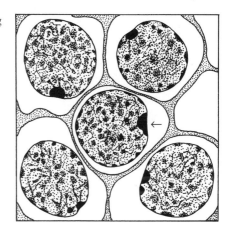

(c) A group of leptotene chromosomes showing each chromosome as a long thin strand with a more compact centromeric region. Some centromeres are fused into a large chromocentre. Remember, each of these chromosomes, like the one indicated by the encircling arrow, consists of two chromatids, even though it may appear to be a single strand.

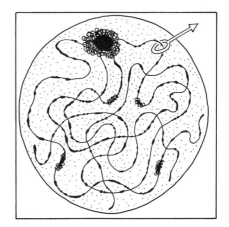

Figure 3.4 (a) A group of nuclei in various stages of **zygotene**. The ends of the chromosomes are all directed towards one side of the nucleus (arrow). In this respect, the nucleus near the centre of the picture is seen in polar view, whilst the one to the right indicated by the arrow is seen in side view. The half of the arrowed nucleus nearest to the arrow is occupied by relatively thick paired chromosomes. On the opposite side, in the region of the chromocentre, the nuclear contents have a much finer texture characteristic of unpaired chromosomes.

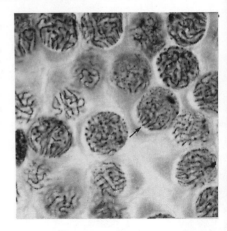

(b) The same features of zygotene nuclei showing the eight ends of the four partially paired bivalents clustered to the side of the nucleus nearest the centrosome. On the opposite side of the nucleus the chromosomes are still unpaired.

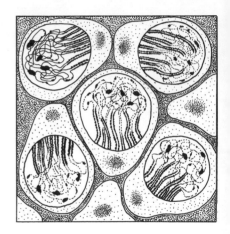

(c) A nucleus with two partially paired bivalents having their four ends attached to the side of the nucleus nearest the centrosome with its two centrioles. At this stage, synaptonemal complexes have formed in the paired regions. Encircling arrow B indicates where two chromosomes have paired to form a bivalent structure. Encircling arrow C indicates an as yet unpaired chromosome *Follow the individual chromosome strands with the point of a pencil from end to end throughout their lengths.*

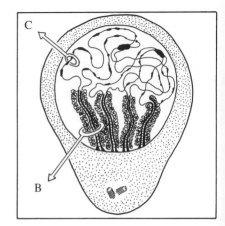

Figure 3.5 (a) A group of **pachytene** nuclei with more condensed chromosomes, now showing as the haploid number of bivalents and still arranged in a bouquet with their ends pointing towards the centrosome outside the nucleus (arrow). These cells were found about halfway along the testis with zygotene and leptotene cells in front of them and later stages of meiosis behind them.

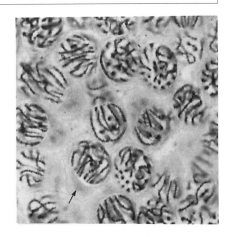

(b) Five pachytene nuclei each with four relatively compact bivalents, ends all pointing towards the centrosome. Remember, each of these structures may appear to be a single strand, but it consists of two synapsed homologous chromosomes, each of which consists of two sister chromatids. The pachytene bivalent is therefore a four-strand structure.

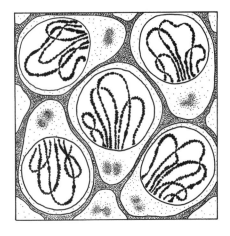

(c) A single pachytene bivalent, signifying the presence of the synaptonemal complex and recombination nodules (arrows). Encircling arrow indicates the entire bivalent assembled around its synaptonemal complex. Insert shows a schematic representation of the relationship between the synaptonemal complex and a recombination nodule (arrow).

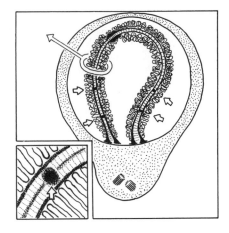

Figure 3.6 (a) **Diffuse diplotene** nuclei. The bouquet arrangement of the chromosomes has now completely disappeared. The ends of the bivalents have detached from the nuclear envelope. The chromosomes themselves have become much less compact so that their axes and contours are no longer discernible. Condensed centromere regions (dense black dots) remain visible but are more numerous, of variable size and more or less randomly dispersed throughout the nucleus.

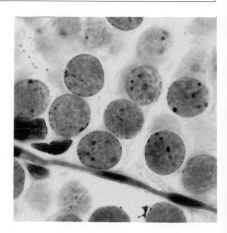

(b) Interpretation of the arrangement of material in nuclei such as those shown in Figure 3.6a.

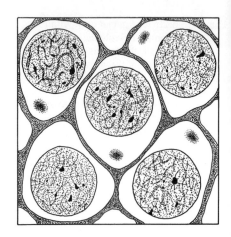

(c) Hypothetical diagram of a diffuse diplotene nucleus with just a single very elongated bivalent in which the component half-bivalents (hb) have separated except in regions where chiasmata (ch) have formed. It should be stressed that the actual form of the chromosomes at this stage has never been observed.

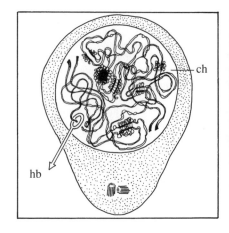

Figure 3.7 (a) Nuclei in **early diplotene**. All chromosomes are now much shorter, more contracted and quite clearly visible. Half-bivalents have separated from one another, remaining attached only where chiasmata have formed. Three places where bivalents can be distinguished are indicated by arrows.

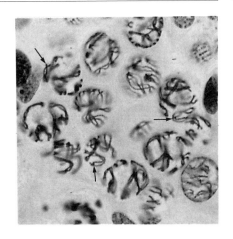

(b) Nuclei in various stages of emergence from a diffuse to a more compact diplotene condition. *Count the bivalents in the nucleus at the top right hand of the picture. Count the chiasmata.*

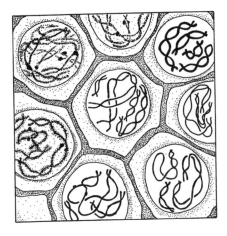

(c) A single diplotene bivalent with four chiasmata (arrows). Insert shows the supposed arrangement of chromatids in relation to localized remnants of the synaptonemal complex in the region of a chiasma (arrow). b, bivalent; hb, half-bivalent.

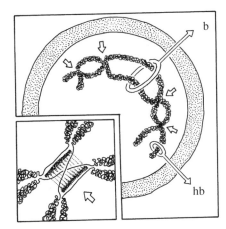

Figure 3.8 (a) Groups of chromosomes in various stages of the first meiotic metaphase and anaphase. A histological section, such as the one from which this picture was made, has depth and it is therefore impossible to bring all the chromosomes in a group into focus in one plane. It is nonetheless clear that the individual bivalents possess the peculiar shapes that are characteristic of first meiotic chromosomes and that there are likely to be about 13 bivalents in each group, which would be correct for this organism. A clearer understanding of the appearance of these bivalents may be obtained by consulting Figure 3.11.

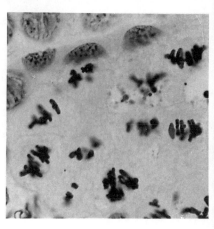

(b) Four nuclei in different stages of first meiotic metaphase and anaphase. The cell in the bottom right is the earlier metaphase; that in the upper right is the more advanced anaphase. Note how the centromeres lead the way to the poles of the spindle. Each chromosome consists of two chromatids. *How many bivalents in the cell at bottom right? How many chromosomes in each daughter cell at top right?*

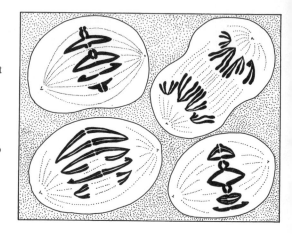

(c) A pair of half-bivalents that have just separated at the earliest point in first meiotic anaphase. Each has two chromatids (Ctd), and in each the sister kinetochores, shown as bars with spindle fibres attached to them, are proceeding side by side to the same pole. Remember, were this a mitosis, these sister kinetochores would be going in opposite directions.

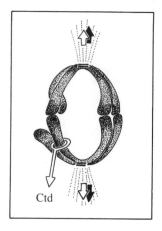

Ctd

Figure 3.9 (a) A group of nuclei in the brief interphase between the first and second meiotic divisions. These nuclei are much smaller than the prophase nuclei (compare with Figure 3.5a) and they contain quite contracted chromosomes each consisting of two chromatids attached only in the region of the centromere. *How many of these chromosomes should there be in each nucleus?* Cells of this kind are found in positions posterior to those in diplotene and first metaphase and often adjacent to the very small nuclei of cells that have completed meiosis altogether.

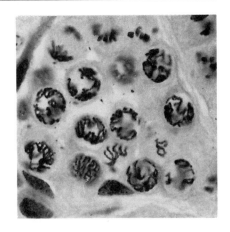

(b) The same type of nuclei as shown in Figure 3.9a. *How many chromatids are represented in each nucleus?*

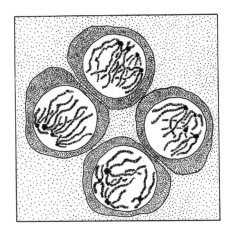

(c) A single chromosome during interphase between first and second meiosis. Two chromatids are held together only at the centromere. At least one of these two chromatids will differ from the ones produced by the S-phase replication that preceded the first meiotic division: why? Arrow encircles just one chromatid (Ctd).

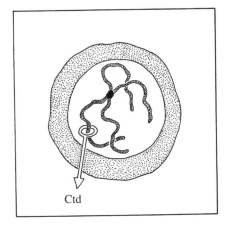

Ctd

Figure 3.10 (a) The images seen here are parts of small clusters of chromosomes, 13 per cluster, in metaphase and anaphase of the second meiotic division. The full number of chromosomes cannot be seen in any one cluster because of the problem of depth of focus. Compare this picture with Figures 3.2a and 3.8a.

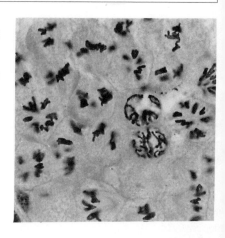

(b) Diagram of six cells in various stages of metaphase, anaphase and telophase of the second meiotic division. *How many chromosomes in the metaphase at the top left? How many chromosomes in each group of the anaphase at bottom middle?* Note that the chromosomes at anaphase are at last single structures, following the separation of the chromatids that were formed a long time previously by replication of the chromosomes in the premeiotic S-phase.

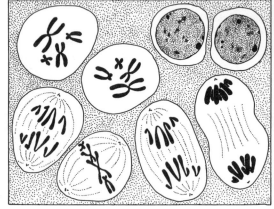

(c) Two single daughter chromosomes, each with a single kinetochore, derived from the two chromatids that have remained associated with one another ever since the very beginning of meiosis. Arrow encircles one single-stranded chromosome (Chr).

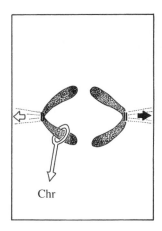

Chr

oocyte lampbrush chromosomes, but not on diffuse spermatocyte chromosomes. Perhaps the most that should be said is that the diffuse and lampbrush stages come at the same point in the meiotic cycle and the spermatocyte situation may represent a rudimentary lampbrush stage that is not allowed its full scope within the narrow confines and time limits of early diplotene in male meiosis.

As the cell nears the end of the diffuse stage, the component halves of the paired chromosomes, so closely associated at pachytene, separate all along their lengths except at the positions where chiasmata have formed and crossing over has taken place (Figure 3.7a, b and c). With advancing **diplotene**, the chromatids produced by the chromosome duplication that took place during the premeiotic S-phase become visible and four-strand bivalents appear. Convention now requires that we call these four-strand structures **meiotic bivalents**. Each bivalent is made up of two **half-bivalents**, and each half-bivalent consists of two **chromatids** (Figure 3.7c).

Diplotene bivalents condense and then the centromeres of each half-bivalent begin to move away from one another, so giving the bivalent a more open appearance (Figures 3.8 and 3.11). The chromosomes are said to be in **prometaphase** as soon as their shape begins to show signs of the tensions set up by early pulling apart of centromeres (Figure 3.11).

As this tension builds up, the bivalents become stretched out parallel to the axis of the developing first meiotic spindle. Half-bivalents are still held together

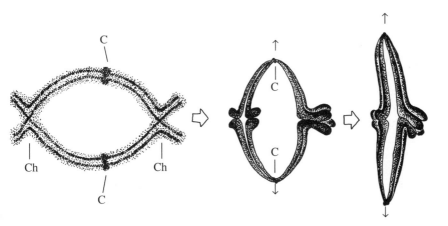

Figure 3.11 The nature of the transition in the form of a bivalent as it reaches the end of diplotene and proceeds into prometaphase and metaphase of the first meiotic division. The change in form is mainly brought about by shortening and compaction of the half-bivalents and pulling of their centromeres towards opposite poles of the division spindle. The half-bivalents remain associated through the chiasmata, which are only pulled apart when the bivalents are tightly stretched across the axis of the cell (solid arrows indicate the directions of pull). C, centromeres; Ch, chiasmata. This chiasmate association serves to maintain the opposing tension between half-bivalents and thus ensures proper co-orientation and disjunction.

at this stage by chiasmata (Figures 3.8a, b and c and 3.11). Note the importance of chiasmata for directing proper distribution of the chromosomes. Separation of half-bivalents at metaphase is prevented by the persistence of chiasmata; when the chiasmata disconnect, abruptly and simultaneously, the half-bivalents spring apart, anaphase movement to opposite poles begins and each cell receives just one representative of each chromosome. If two homologous chromosomes fail to establish chiasmate association with one another, then they will associate with the spindle independently. Half the time they will, by chance, orientate towards the same pole, and all the gametes resulting from such a mis-segregation (formally referred to as **non-disjunction**) will be **aneuploid** (Greek: *a*, without; *eu*, well; *aploos,* onefold; without good onefoldness, having more or less than the normal haploid chromosome number): half of them will have no representatives of the chromosome, while the other half will have two. Aneuploidy of this kind is nearly always lethal. Consequently, for reasons that are quite distinct from the genetic advantages of crossing over and recombination, it is in an organism's best interests to maintain a meiotic recombination system that guarantees at least one cross-over event on each chromosome pair, just to ensure the proper distribution of the chromosomes at anaphase 1.

As the cell progresses from **metaphase** through to **anaphase** (Figure 3.8b), each bivalent is dismembered into two half-bivalents, one going to each pole of the first meiotic spindle (Figure 3.8c). The 'cutting of the pack' and the 'shuffling of the cards' have both been accomplished. The first meiotic division is over and chromosomes that were once described as 'homologous', but have since been reorganized as a result of crossing over, have now been equally but randomly separated from one another and redistributed into two new cells.

Prophase of the second meiotic division follows almost immediately. The chromosomes never fully decondense after first meiosis. The component chromatids come apart all along their lengths except at the centromeres (Figure 3.9a, b and c). The chromosomes recondense and quickly proceed through a metaphase and anaphase that are basically similar to a mitosis but quite different, of course, in the sense that only the haploid number of chromosomes is involved (Figure 3.10a, b and c). Accordingly, in Figure 3.10a each group of chromosomes is certainly made up of 13 rather than the 26 that we saw at the outset, and in Figure 3.10b each metaphase group consists of four chromosomes, each of which has two chromatids.

It may be helpful to deal now with two very basic practical matters that are of real importance when looking at a microscope preparation with a wide range of meiotic stages. The first relates to the problem of distinguishing between a mitotic metaphase, a first meiotic metaphase and a second meiotic metaphase. It is really quite simple, provided one knows the approximate haploid chromosome number for the species in question.

- A cell in **mitotic metaphase** will show the full diploid number of chromosomes and each chromosome will be made up of two strands or chromatids. In humans, for example, there will be 46 chromosomes on the spindle and each will obviously consist of two chromatids.

- A cell in **first meiotic metaphase** will have the haploid number of paired chromosomes, bivalents, and each of these will have one or other of the strange shapes that result from synapsis, chiasma formation and the events of diplotene and prometaphase.
- A cell in **second meiotic metaphase** will show the haploid number of chromosomes (in humans, 23 as compared with 46), each of which will have a simple form and consist of two chromatids.

The problem, then, becomes one of making a guess as to how many chromosomes can be seen and then deciding on whether or not we are looking at half-bivalents joined by chiasmata.

The second matter relates to the relative abundance of cells in the different stages. Basically, the greater the time occupied by a stage, the more common will cells in that stage be. Accordingly, most preparations of testis show many cells in prophase of the first meiotic division and very few in second meiotic prophase. The first may occupy many days, the second lasts only a few hours (Figure 3.1).

Meiosis is remarkable for its timing and its cycles of condensation and decondensation. It begins with a long critical introduction that is noted for its consequences but exceedingly hard to analyse from the cytological standpoint. It ends with a burst of activity, from diplotene through to second anaphase when the chromosomes are almost continuously moving and changing form. During this cycle, the chromosomes at one time or another pass through almost every conceivable state of condensation and decondensation, assuming highly distinctive and supposedly functional forms at each specific stage. What does it all signify? And, of course, this takes no account of the variability that exists from one organism to another. We should note, for example, that not all amphibians have a diffuse diplotene stage. Some organisms have no bouquet arrangement of chromosomes in their first meiotic prophase. Others have chromosomes stretched diametrically across the nucleus. Some start the pairing process at the ends of the chromosomes, others in the middles and yet others at apparently random points all along the lengths of the chromosomes. This kind of variability is the very essence of the 'comparative approach' in biology. It represents nature's experiments, tested by natural selection over hundreds of generations, some successful and some not, sifting out the truly significant and important features of a process from the frills and trivia.

THE SYNAPTONEMAL COMPLEX

Besides the formal features of meiosis that we can see with a light microscope, there is one more matter that needs to be dealt with before we can focus on the main problem areas. During synapsis, homologous chromosomes are not directly in contact with one another but are slightly separated by a narrow space occupied by a structure known as the **synaptonemal complex** (SC), first described about 40 years ago in sections of spermatocyte nuclei from crayfish. An SC is mainly composed of protein and it essentially represents a structural differentiation of the boundary zone between two synapsed half-bivalents. An electron micrograph and a diagram of the complex in longitudinal section are shown in Figures 3.12 and 3.13.

An SC is flat or ribbon-like in cross-section and is easy to see in electron micrographs of thin slices (sections) of cells in pachytene. One interesting approach in cell biology has been to reconstruct the intranuclear arrangement of chromosomes from the distribution of SCs in a complete and continuous series of sections through a pachytene nucleus. About 20 years ago, a very simple technique was developed for spreading the contents of spermatocyte nuclei on an air–water interface, picking them up on an electron microscope specimen holder (an EM 'grid') and visualizing by electron microscopy SCs from which most of the chromatin had been removed. Each complex appears as a long, thin filamentous structure whose structure and general features are such as would be

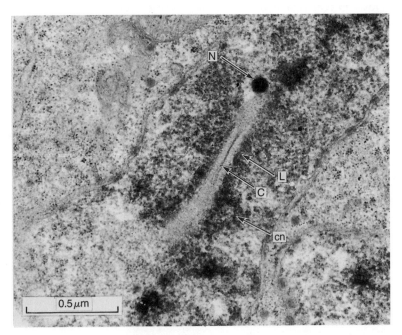

Figure 3.12 An electron micrograph of part of a section through a cell in prophase of meiosis. Part of a synaptonemal complex has been sectioned longitudinally and shows the lateral (L) and central (C) elements, the surrounding mass of chromatin (cr), and a single recombination nodule (N). (Picture kindly supplied by Dr Gareth Jones of the University of Birmingham.)

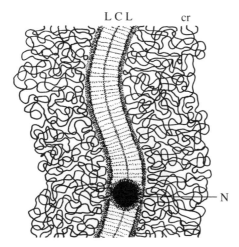

Figure 3.13 Representation of a synaptonemal complex viewed in longitudinal section, as in Figure 3.12. Remember, each of the chromatin masses that lie immediately outside the lateral elements consists of two chromatids. C, central element; L, lateral element; cr, chromatin; N, recombination nodule.

expected on the basis of what is already known from sectioned material (Figures 3.14 and 3.15).

Each SC usually has a local differentiation that marks the position of the centromere, and the relative lengths, arm ratios and centromere indices correspond closely with those of the mitotic metaphase chromosomes from the same species. Just exactly what the SC is has still not been clearly established. It is not the chromosome, but rather something that is added to the chromosome during meiotic prophase. Nonetheless, it is very like the chromosome in several important respects. It has the right length. Its centromere is in the right place. It is single before synapsis and double after synapsis. It is a side-by-side structure. Therefore it provides us, through analysis of serial sections or surface spreads, with excellent opportunities to trace the behaviour of meiotic chromosomes during the synaptic stages of prophase, and that is something that would not otherwise be possible.

As to the actual function of the SC, that remains one of the many interesting unknowns in the meiotic story. During pairing and subsequently throughout pachytene small dense generally spherical structures are associated with SCs. They are called '**recombination nodules**' (RN) (Figures 3.12 and 3.13). They were formally discovered and reported in 1975, but they are so structurally unspectacular that they were probably thought to be pieces of 'dirt' on electron micrographs for years before that. The interesting thing about them is that they always associate with SCs and their positions and numbers in mid- to late pachytene correspond closely with genetic cross-over events. There are two kinds of them appearing at different times between the beginning and end of

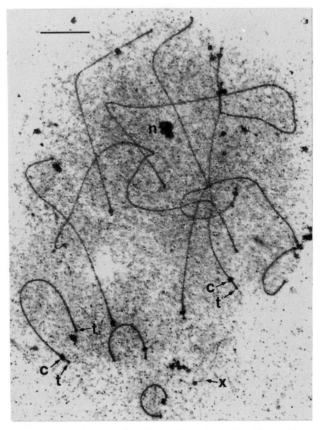

Figure 3.14 A photograph taken with a light microscope of a silver-stained, surface-spread pachytene nucleus from *Locusta migratoria*, showing 11 autosomal synaptonemal complexes and traces of the univalent X axial core (X). c, centromeres; t, telomeres; n, fragmented detached nucleolus. The scale bar in this picture represents 10 μm. (Picture kindly provided by Dr Gareth Jones of the University of Birmingham and reproduced with the permission of John Wiley & Sons from Macgregor, H.C. and Varley, J.M., *Working with Animal Chromosomes,* 2nd edn, 1988.).

synapsis. They almost certainly have some role, together with the SC, in bringing homologous chromosomes together in register with one another and in promoting the molecular events that lead to cross-overs. The matter has been well reviewed by Carpenter (1987).

CHROMOSOME PAIRING

Against that background of form and behaviour, let us now consider the first of our three problem areas: chromosome pairing. To begin with it is perhaps as

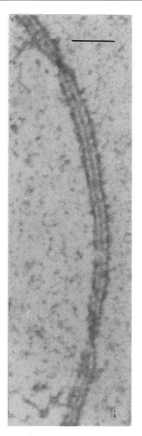

Figure 3.15 An electron micrograph of a portion of a surface-spread synaptonemal complex clearly showing its tripartite structure. The scale bar represents 0.5 μm. (Picture kindly provided by Dr Gareth Jones of the University of Birmingham and reproduced with the permission of John Wiley & Sons from Macgregor, H.C. and Varley, J.M., *Working with Animal Chromosomes,* 2nd edn, 1988.).

well to admit that, despite years of painstaking research, we really know remarkably little about pairing. It is a phenomenon that is still waiting to be explored and explained, a challenge to future generations of investigators who are aware of the problems, can formulate the right questions and combine to their utmost limits the skills of modern microscopy, biotechnology and the comparative approach.

Meiotic pairing that is followed by recombination must consist of at least two distinct steps, namely chromosome pairing and precise matching of DNA base sequences. It usually involves homologous chromosomes. However, it can, and frequently does, at first involve any two chromosomes that come within less than half a micrometre of one another in a meiotic prophase nucleus. The SC that characterizes paired segments is actually quite indifferent to homology. The fact that pairing normally happens between homologues in approximate alignment with one another has led to a strong suggestion, for which there is ample evidence, that homologues presort themselves, i.e. undergo some kind of presynaptic alignment so that when zygotene commences the likelihood is that most pairing will be homologous. It has also been suggested that another unknown mechanism must exist that acts independently of the SC in recognizing and bringing homologues together. Only after the preselection of homologous chromosomes and their strong and close connection by the SC can precise matching of DNA molecules take place. Accordingly, the SC is seen not as a primary contact between homologues but rather as a mechanical framework within which the molecular processes of sequence matching and recombination can take place.

Where does pairing normally begin? It may start in the region of centromeres, as in some salamanders, where there is a clumping of the regions of the chromosomes that flank the centromeres, the centromeric heterochromatin, into a chromocentre at leptotene and zygotene (Figures 3.3 and 3.4), or it may begin at the ends of the chromosomes where they attach to the nuclear envelope. However, it is quite clear that although chromocentres and nuclear membrane attachments do facilitate the initiation of synapsis, the process can also begin in other places. In *Schistocerca gregaria,* the desert locust, occasional tetraploid spermatocytes have four of each chromosome and quadrivalents form in which the complex patterns of pairing can only be explained by supposing that pairing starts at at least six independent sites along the length of the chromosome. Structural rearrangements provide us with more clues. Pairing has been examined in organisms that are heterozygous for insertional translocations in which a piece of one chromosome is required to pair up with its homologous counterpart that has been excised from its normal site and inserted into the middle of another non-homologous chromosome arm. In such cases, chromosome ends or centromeres are of no help in bringing the two homologous segments together. Indeed, the complex patterns of pairing that are found in some translocation and inversion heterozygotes suggest strongly that some positive recognition forces must be at work in a zygotene nucleus, but we have absolutely no understanding of the nature of these forces nor has the scale of the problem been properly assessed.

Next we can usefully ask certain questions about nuclear envelope attachments. When do they form? The answer is not entirely clear. What is seen is the association of the end of an SC or a single axial complex with the inner surface of the nuclear envelope. Is the chromosome attached to the nuclear envelope before the components of the axial or synaptonemal complexes are assembled? Are chromosome/nuclear envelope attachments positioned randomly over the nuclear envelope? Again the answers are unclear. In those cases where the ends of the chromosomes have been mapped from serial sections at a stage when they are widely distributed over the nuclear envelope, the relative positions of specific attachments are not regular. Where attachments have been mapped in nuclei where the ends of the chromosomes are grouped towards one pole of the nucleus, homologous axial complexes are close together and SCs are soon formed. In such cases it is impossible to decide whether the chromosomes have been drawn together by the formation of the SC or simply encouraged to form complexes by the juxtaposition of their ends.

Do the nuclear envelope attachments move? Here we can make some more definite statements, though still without providing any really satisfying answers. Where SCs form near to but not right at the point of attachment, the distance between the attachment points is usually appreciably greater than the distance separating the two halves of the SC; yet eventually the SC extends all the way to the nuclear envelope. So attachment points must be capable of moving over the short distances that are involved in drawing them together for completion of the SC. More convincing evidence comes from observations on locusts, spiders and salamanders. In *Locusta*, the ends of the chromosomes all point towards the centrioles that lie just outside the nucleus. At pachytene, two pairs of centrioles move apart until they lie on opposite sides of the nucleus. As this happens, some of the attached chromosome ends follow one pair of centrioles and the others remain stationary in the relative sense. Here we have good evidence of movement of attachment site, albeit after pairing has been completed. The same kind of event takes place in spider spermatocytes, where it culminates in a stage that has been described as 'pseudo-anaphase', so called because two more or less equal groups of chromosomes come to lie on either side of the nucleus, only to disperse again at diplotene before condensing for the real metaphase and anaphase. In salamander spermatocytes, axial complexes are associated with the nuclear envelope in late leptotene and early zygotene, before the bouquet arrangement of the chromosomes has formed, suggesting that attachments form and then migrate to one side of the nucleus to assemble into the base of the bouquet. Why do attachment points associate with the centriolar complex (Figure 3.4)? Are there structural connections between the two? Does extensive movement of attachment points involve synthesis of nuclear envelope? Exactly what is moving with respect to what? What happens at the end of pachytene when the bouquet arrangement suddenly relaxes and the chromosomes detach from the nuclear envelope?

Another important aspect of pairing is its timing. Does it all happen at once? Is it programmed? Does pairing between two chromosomes happen at a constant rate? This is an area where the technique of spreading SCs on an air–water interface has helped enormously. Suffice to say here that in a normal diploid karyotype there is no special programme of pairing and the rate at which pairing proceeds and is completed is not a function of chromosome length. In sala-manders, the situation seems quite uniform in the sense that a wave of synapsis passes slowly across the nucleus from the base to the apex of the bouquet, affecting each of the metacentric chromosomes at about the same rate.

Closely connected with the matter of timing is that of completeness. In several cases that have been examined it is known that chromosomes do not pair all along their lengths. The grasshopper *Stethophyma grossum,* for example, has 11 telocentric autosomes. The three shortest of these pair completely. The remaining eight show varying degrees of partial pairing, with one of the larger chromosomes pairing over less than half its length.

Two more matters need to be mentioned because they give us a small glimpse of some of the molecular complexities that remain to be worked out. In some of the classical literature on meiosis, there is concern about interlocking. If we start with a spherical chamber occupied by a random arrangement of 50 or so long strands of material for each of which there is an identical or homologous partner, and we set about zipping the partners together all along their lengths, starting at both ends of each pair, it would seem to be almost inevitable that some pairs will become interlocked or tangled with one another (Figure 3.16). It would be a

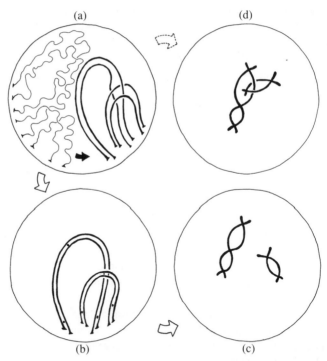

Figure 3.16 Opportunities for interlocking between bivalents during synapsis and the potential consequences at diplotene. On the left side of the nucleus shown in (a) leptotene chromosomes are attaching their ends to the nuclear envelope and preparing to commence synapsis. A little later, when synapsis is well advanced, we see that two pairs of homologous chromosomes have become interlocked. If the interlock is not resolved, then the result will be an irrevocable interlocking of two diplotene bivalents such as is shown in (d), which will have serious consequences at anaphase of meiosis 1. In the event, the interlock is resolved before diplotene and no interlocking of diplotene bivalents is seen, (b) and (c).

miracle if they did not. Nevertheless, when our meiotic nucleus emerges from pachytene and the bivalents become clearly distinguishable at diplotene, there are no interlocks or tangles.

At one time it was thought that interlocks were rare accidents and considerable effort went into demonstrating how their frequency could be increased by various physical and chemical treatments. Nuclear envelope attachment was considered important. After all, if chromosomes are not already entangled, then by tacking down their ends we eliminate all further chance of entanglement. The problem disappeared, only to be replaced by a much greater one, when it was clearly demonstrated in a fungus, an insect and man that interlocks were commonplace in late zygotene nuclei, but all of them disappeared before the end of pachytene. Resolution of the interlocks and repair of broken ends require breakage and precise rejoining of DNA strands from one or both sister chromatids in a half-bivalent. Now it is one thing to find the molecular machinery that can do this kind of task. It is quite another thing to understand how that machinery identifies an interlock and effects the disentanglement.

Related to the matter of interlocking is the problem of bivalent formation in polyploid cells. In a tetraploid nucleus there is the potential to form quadrivalents, trivalents, bivalents and univalents, and all of these were identified in tetraploid oocytes of the silkworm, *Bombyx mori*. However, at metaphase 1, only bivalents were observed. The same consideration applies to wheat (*Triticum aestivum*), a plant that is known as an **allohexaploid** with a karyotype made up of six sets of chromosomes derived from three separate species. It arose, probably in neolithic times (*c.* 10 000 years ago), as a natural cross between a tetraploid and a diploid species. The possibilities for meiotic synapsis in such an organism are endless, and indeed three-dimensional reconstructions of wheat nuclei at zygotene reveal multivalents, interlocking, pairing between non-homologous chromosomes and all manner of unruly events. Yet reconstructions of pachytene nuclei reveal only properly paired and wholly respectable bivalents. What is more, crossing over and chiasma formation in wheat is delayed until completion of the pairing correction process.

Naturally enough, a whole range of experiments have been carried out with wheat because of its value as a food crop. Of particular significance in the present context is the **Ph gene** complex on the chromosome designated 5B in wheat. In the wheat karyotype, pairs of chromosomes that stem from one of the parent species are referred to as homologous, whereas chromosomes of corresponding number but from different parent species are called **homoeologous**. The Ph complex decides whether pairing will be strictly homologous or a free-for-all between homoeologues. High doses of Ph suppress pairing altogether, even that between homologous chromosomes. Genes are known that affect the pairing of just one chromosome in the set. Others are known to have their effects during or soon after the last premeiotic mitosis and yet others are known to modify the effects of the Ph locus, so that different allelic combinations of them will generate a series of conditions ranging from total asynapsis to full homoeologous pairing and the free formation of multivalents that persist through to anaphase 1.

For the most part, the molecular biology of meiotic pairing remains a mystery and a challenge, but perhaps not for long. Hitherto most investigators have coupled microscopy with a study of carefully selected organisms that offer special opportunities for meiotic studies. More recently, new molecular techniques have been developed that can be combined with light or electron microscopy to solve molecular problems: for example, immunofluorescence and confocal microscopy are sure to make an impact, just so long as the questions are clear and the right organisms are chosen for study. One of the highest priorities must surely be the mechanisms for primary recognition and pairing together with those that sort out the mistakes and do the tidying up during late zygotene and pachytene.

CHIASMATA AND CROSSING OVER

Pachytene is followed by diplotene, during which we see the relaxation of pairing forces and the separation of half-bivalents until they are held together only by the chiasmata. That chiasmata are points of genetic exchange between homologous chromatids is plain to see in good cytological preparations where two of the four chromatids in a diplotene bivalent literally cross over (Figure 3.17). It was also conclusively proved in the classical experiment carried out by Creighton and McClintock in 1931 in which the inheritance of cytologically visible chromosome markers was examined along with genes occupying known positions on the same chromosomes as the cytological markers. When the genes crossed over, as determined by expression in F1 phenotypes, so did the chromosome markers.

The basic rules of chiasma formation are as follows:

1. Crossing over is a consequence of exchange between homologous chromatids.
2. Only two out of the four chromatids are involved in each cross-over.
3. Crossing over is reciprocal and symmetrical in the sense that each cross-over gives a tetrad containing two parental non-cross-over chromatids and two complementary cross-over chromatids.
4. Chiasmata form after premeiotic replication and after chromosome pairing at first meiotic prophase.
5. Crossing over is associated with a phenomenon that we call **positive interference**, which means that the occurrence of one cross-over in a chromosome arm reduces the chances of another cross-over occurring in the same arm.
6. Crossing over is associated with another phenomenon known as **negative**

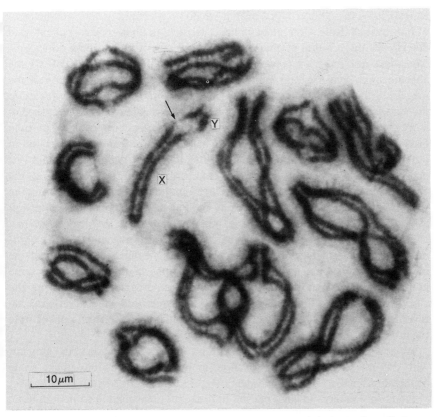

Figure 3.17 A set of diplotene chromosomes from the testis of a salamander called *Oedipina uniformis*. There are 13 bivalents altogether, one of which is an XY pair of sex chromosomes that are tenuously connected by a terminal chiasma (arrow). The four-strand structure of the bivalents is clear in this photograph and the actual crossing over of chromatids from one half-bivalent to the other at chiasmata is particularly well displayed. (Picture kindly provided by Dr James Kezer of the University of Oregon and reproduced with the permission of Macmillan, New York, from Novitski, E., *Human Genetics*, 1977, and with the permission of John Wiley & Sons from Macgregor, H.C. and Varley, J.M., *Working with Animal Chromosomes*, 2nd edn, 1988.)

chromatid interference, which is to say that involvement of two chromatids in a cross-over in no way reduces the chances of the same two chromatids being involved in another cross-over at another point along the chromosome. This proved to be a necessary concept, since it was at one time assumed that there would be *positive* chromatid interference.

7. Chiasmata or cross-overs are never random in position. Besides the interference factor, the positions and numbers of chiasmata are characteristic for each species and sometimes even for each of the two different sexes within a species.

Rules 1 to 6 are matters of fact, even if we do not yet fully understand the basis of interference. Rule 7, on the other hand, leads us to think seriously about what determines the position of chiasmata and how they form. In some organisms, chiasmata are more or less evenly distributed along the arms of all the chromosomes and the total number of chiasmata is largely a function of the lengths of the chromosomes' arms. Such appears to be the case with regard to the very large chromosomes of some plethodontid salamanders, where three or more chiasmata can form in arms of some of the longer chromosomes. In some organisms, the distribution of chiasmata may be the same in both sexes, but in others it may be quite different. The grasshopper *Stethophyma grossum* is a good example. Here, one or two chiasmata form near the centromeres (all the chromosomes of this animal are acrocentric) in the male, and none form in the greater part of the chromosome arms. In the female, most chiasmata form well away from the centromeres. Likewise, in the palmate newt, *Triturus helveticus*, chiasma distribution is strikingly **proterminal** (near the ends of the chromosomes) in the male, with all the chiasmata right out near the ends of the metacentric or submetacentric chromosomes; but in the female, the same number of chiasmata are found occupying intercalary positions in the chromosome arms. In the crested newt, *Triturus cristatus*, the situation is the reverse, with liberal scattering of chiasmata in the male and strong **procentric** (near the centromeres) localization in the female.

What is the basis of chiasma localization? The best we can do at the moment with this question is to focus on three more specific questions, make some observations and do some experiments.

First, do you have to have pairing in order for chiasmata to form and, if so, then might chiasma localization be a direct consequence of specifically limited meiotic pairing? Although in most cases the evidence points towards more or less complete pairing of all chromosomes, it has been convincingly shown in *Stethophyma* that in regions where there are no chiasmata there is no pairing of the chromosomes. This, and other evidence from other animals, including another grasshopper, a flatworm, a fungus and man, strongly indicates that two of the prerequisites for chiasma formation are meiotic pairing and the formation of a synaptonemal complex.

Second, is there anything that we can actually see in the regions where chiasmata form? Careful examination of serial sections of spermatocytes has shown a good relationship between chiasmata and recombination nodules (RNs). During zygotene, synaptonemal complexes form and a large number of small round RNs are scattered along their lengths. Through the early part of pachytene, the RNs decrease in number and those that remain become much larger and spread across the entire width of the SC to form a connecting bar. Towards the end of pachytene, most of the RNs have disappeared. About as many remain as there are likely to be chiasmata at diplotene, and those that remain all have the bar-like appearance. So RNs do, as their name suggests, have something to do with recombination, though precisely how they work has yet to be discovered. An interpretation of their relationship to the SC and the chromatin is shown in Figure 3.18.

Can we explain interference on the basis of the behaviour of RNs? Once again, the best we can do here is a hypothesis based on the observed facts. We know that there are preferred 'domains' in individual chromosome arms within which chiasmata form. The hypothesis is as follows. There are a limited number of RNs. A domain in an SC has a high affinity for RNs in its centre and a low affinity towards its ends. The occurrence of a cross-over in a domain reduces the probability of attachment of new nodules and causes those nodules already there and not yet involved in crossing over to be released from that domain.

Two sources for positive chiasma interference are therefore identified: (1) limited availability of RNs restricts the number of cross-overs that can take place in a chromosome arm; and (2) preferential attachment of RNs to certain domains of the SC, coupled with a rejection mechanism that redirects RNs after the end of zygotene to domains devoid of nodules, will reduce the chances of two closely positioned cross-overs.

It takes only a little perception to realize that we have really not made a lot of progress. All we have done is to exchange the word interference for words like 'preferential' and 'rejection' and we have brought SCs and RNs into the picture.

Finally, in relation to crossing over, we know, of course, that it must involve breakage and repair of the two DNA duplexes that represent the two chromatids that are involved in the cross-over. If RNs do indeed effect recombination, then

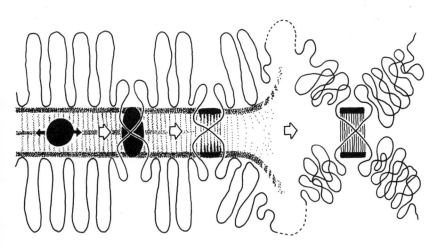

Figure 3.18 A synaptonemal complex showing the various stages, from left to right, in the involvement of a recombination nodule in crossing over. On the extreme right of the diagram is the cross-over as it might exist in very early diplotene, after disappearance of the synaptonemal complex. Note in particular: only one chromatid per half-bivalent is shown in this diagram. For the sake of clarity, the non-cross-over chromatids are not shown.

we might reasonably expect DNA repair synthesis to take place in their vicinity during late zygotene and pachytene, which indeed it appears to do. An experiment employing [3H]thymidine incorporation, autoradiography and electron microscopy has been carried out, with considerable skill it should be said, since the question being addressed demands the utmost efficiency from all the techniques involved. Oocytes from *Drosophila* were incubated in [3H]thymidine, then fixed, embedded in plastic and sectioned, thickly for light microscopy and more thinly for electron microscopy. The sections were covered with a radiation-sensitive (nuclear track) emulsion and left to 'expose' in darkness for varying periods of 5 to 135 days. The preparations were then processed, stained and examined to see how silver grains produced in emulsion overlying local concentrations of [3H]DNA were distributed in relation to SCs and RNs. The results were quite conclusive. Silver grains were 2.6 times more common in the vicinity of large and bar-shaped RNs than would be expected if they represented background labelling not produced by exposure to [3H]DNA. The result is fairly regarded as evidence of repair-type synthesis of DNA at the site of RNs, and it gives support to the notion that RNs are part of the mechanism that produces reciprocal exchanges.

Everything that I have said so far about meiosis adds up to very little in terms of hard fact and clear understanding. That is scarcely surprising. The process is poorly understood because it is not at all easy to study. Gonads and germinal tissues generally do not lend themselves well to simple analytical biochemical approaches and, even from the cytologist's viewpoint, they are complex organs that need the subtlest and most devious of experimental approaches. Nevertheless, it remains a challenging field and one in which experimental ingenuity has repeatedly paid handsome dividends and in which the right choice of experimental material, essentially the full exploitation of meiotic variability, is of crucial importance for success.

DISJUNCTION

The last meiotic matter that we have to consider is **disjunction**, a word that means the disjoining or separation of half-bivalents at first meiotic anaphase, such that one goes to one spindle pole and the other to the opposite pole. It is remarkable in two respects. First, it clearly cannot happen with chromosomes that have failed to form chiasmate associations. Univalents have nothing from which to disjoin! They can only behave independently and randomly, which augurs badly for a balanced distribution of chromosomes at first anaphase. Secondly, disjunction involves chromosomes that are composed of two sister chromatids whose centromeres in any other mitotic division would normally separate and go to opposite poles, an event known as **equational segregation**. In first meiosis, sister centromeres stay together and homologous centromeres disjoin, giving rise to a **reductional segregation** (Figure 3.19).

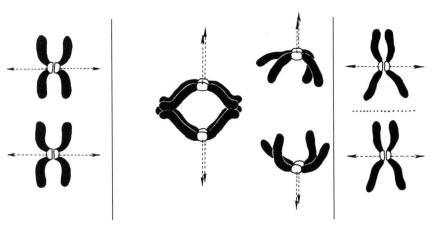

Figure 3.19 The differential behaviour of kinetochores at mitosis (extreme left), first meiotic metaphase and anaphase (centre) and second meiotic metaphase (right). In mitosis and second meiotic anaphase, sister kinetochores, belonging to sister chromatids, separate. In first meiotic anaphase sister kinetochores remain together and move side by side to the same spindle pole.

To explain disjunction we have to understand **kinetochores**, the structures that we find associated with the centromeric region of the chromosome from metaphase onwards. We know about their structure. We know more and more every day about their proteins and how they relate, on one side, to DNA sequences in the centromeric region of the chromosome and, on the other side, to microtubules connecting the chromosome to the spindle pole. Most importantly, we know that the kinetochore is a compound plate-like structure that forms on the side of the chromosome or chromatid that faces towards the nearest spindle pole (Figure 3.20).

Immediately before the start of anaphase in a mitotic division, sister kinetochores lie opposite one another and face towards different poles. At anaphase they go in different directions and the chromosomes segregate equationally. In first meiotic metaphase, sister kinetochores lie next to one another, facing in the same direction, and the chromosomes segregate reductionally, with both sister kinetochores on the same half-bivalent going the same way. Therefore we may suppose that the direction in which a kinetochore faces decides the way it will go at anaphase. But, of course, there is more to it than that! Why, you will naturally ask, *do* sister kinetochores face in different directions in mitosis and in the same direction in first meiosis?

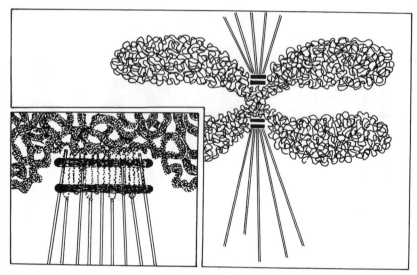

Figure 3.20 The arrangement of the principal components of a kinetochore on a eukaryotic metaphase chromosome. Each chromatid has a kinetochore positioned on the side that faces towards the nearest spindle pole. The kinetochore is a plate-like structure that consists of three layers, two that appear dark (electron dense) in side section view, with a lighter, less electron dense layer between them. The inset shows the arrangement of the kinetochore layers in relation to spindle and chromatin fibres. The upper part of the picture is occupied by chromatin. Nine spindle microtubules enter the picture at the bottom; these penetrate the kinetochore to various depths. Some stop at the outer surface of the kinetochore; others pass right through to the chromatin beneath the inner kinetochore layer. Some chromatin fibres appear to penetrate the inner kinetochore layer and are present in the middle and outer layers.

Let us look at some experiments that may help us to understand the problems of disjunction and assess the forces and constraints that are at work. The experiments are really quite remarkable in the sense that they involve actual micromanipulation – movement by *hand* – of chromosomes in living cells while at the same time observing and photographing them with the help of phase-contrast microscopy. In practice, a minute glass microhook is introduced into a cell during metaphase or anaphase of first meiosis and moved by the operator, working through a micromanipulator which greatly scales down and damps the natural movements of the hand, in such a way as to change the position and orientation of individual chromosomes. The cell is then kept under constant observation and photographed in the plane of the displaced chromosomes at regular short intervals.

If we bend a bivalent at the beginning of metaphase, such that both its pairs of sister kinetochores face towards the same pole (Figure 3.21a), then it will immediately bend back again and reorientate itself in the proper way, which is to say that by the time we get into metaphase each pair of centromeres and their associated kinetochores seems committed to go in a particular direction. However, if we manipulate a bivalent that consists of two acrocentric chromosomes into a U formation at early metaphase of first meiosis, so that both its pairs of sister kinetochores point towards the same pole and we hold it that way by leaving the microhook at the base of the U and applying a slight pull towards the opposite spindle pole (Figure 3.21b), then the bivalent will not reorientate and both pairs of sister kinetochores will ultimately go to the same pole.

Two other experiments of the same general kind are of interest. First, if we

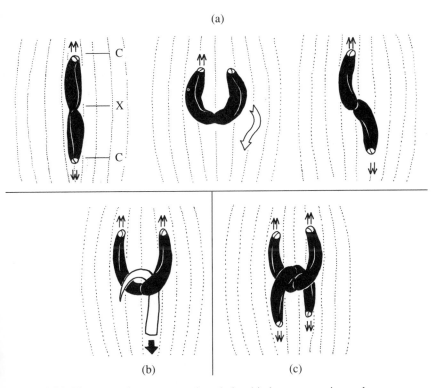

(a)

(b)   (c)

Figure 3.21 Three experiments to test the relationship between tension and co-orientation stability in first meiotic metaphase/anaphase bivalents. In all three cases the chromosomes are telocentric (C, centromeres) and the bivalents are maintained by a single near-terminal chiasma (X). (a) A bivalent is bent so that the centromere regions of both half-bivalents point towards the same pole. It soon reorientates so that its centromere regions point to opposite poles. (b) A bivalent is bent as in (a) and then tension is applied to the base of the U through a microhook. No reorientation takes place and both centromere regions tend to orientate towards the same pole, away from the direction of tension. (c) Two bivalents are bent and hooked around each other, each with both their centromere regions pointing towards the same pole. No reorientation takes place because tension is maintained by each bivalent pulling against the other.

cool a male grasshopper down to 10°C for 10 days and then heat him up again to 25°C for 2 hours and then cool him again to 10°C for about 5 hours, subsequent examination of live spermatocytes at room temperature shows extensive instability in centromeric orientation at first meiotic metaphase. The centromeres of some bivalents wander indecisively between the two spindle poles before finally orientating themselves for a proper reductional disjunction. This temperature-induced instability can be checked by catching a bivalent with the microhook and applying tension to it. The centromeres then stabilize in such a way as to oppose the tension. Release the tension and they again become indecisive and unstable. Reapply it and they reorientate.

The second experiment sets out to make both pairs of sister centromeres go to the same pole by micromanipulating two bivalents into U formations and then hooking the Us around each other (Figure 3.21c). Consequently, two pairs of sister centromeres are made to pull in the same direction against another two pairs. The experiment was entirely successful and illustrates very clearly the importance of physical tension, opposing force, in maintaining metaphase 1 orientation stability.

Two final points are important in this context. First, there can be no doubt whatever that the centromeres of meiotic half-bivalents are double structures. In other words, we cannot explain reductional disjunction by supposing that meiotic centromeres are single and mitotic ones are double. Secondly, it seems quite certain that there is a change in the properties of the sister centromere pair immediately after the first meiotic division. This much we know from the following experiment, again involving some very clever micromanipulation. Two grasshopper spermatocytes, one in first meiotic metaphase (1MM) and the other in second meiotic metaphase (2MM) were fused together such that two spindles were in the same cell. A chromosome was then detached by micromanipulation from the 2MM spindle and put near to the 1MM spindle. The sister centromeres then became attached through their kinetochores to the 1MM spindle and the sister chromatids separated and moved to opposite poles along with the meiotic half-bivalents. This behaviour is different from that of an unpaired chromosome which, having failed to establish chiasmate association with its homologue, then becomes attached to the 1MM spindle in the normal course of events. Such a chromosome is usually orientated towards just one pole and both its chromatids go towards that pole at first meiotic anaphase.

There are not many hard conclusions that we can reach on the basis of these few facts and observations, but there are some quite clear questions that we can usefully turn over in our minds. What determines which way a pair of centromeres and their associated kinetochores will face? How does tension work in stabilizing the behaviour of centromeres? Could it be that movement, however slight, in one direction within the framework of the spindle, will stimulate the growth of microtubules away from the outer face of the kinetochore in the opposite direction? Such movement would be produced by micromanipulation or, in normal circumstances, it could, for example, be a consequence of the final stages of chromosome condensation and contraction during first meiotic metaphase.

From the standpoint of chromosome evolution and in so far as we would like to be able to understand the troublesome matter of meiotic non-disjunction, we need to know why two sister chromatids may sometimes pass to the same pole of the mitotic spindle and why meiotic chromosomes that have failed to form chiasmate associations with their homologues misbehave at first meiotic anaphase.

DURATION OF MEIOSIS

To round off our discussions about meiosis it seems appropriate to think for a moment about its duration, since of all the features of the process this is the most variable. The time required for the whole sequence of events from the start of leptotene to the end of second anaphase varies from a few hours in yeast to several weeks in man and some other animals. If we take account of the developmental holds such as occur at diplotene in many animals, then of course meiosis in the human female, for example, can last for as long as 50 years.

In general terms, we can say that differences in the duration of meiosis from one organism to another reflect differences in all stages. Premeiotic S, G2, prophase, metaphase and anaphase are all variable and, as a rule, they are all much longer than the corresponding stages associated with mitosis in somatic cells.

The duration of meiosis is usually quite constant for a given species, as are the durations of individual stages. The Chinese hamster, for example, spends 75% of its meiotic cycle in pachytene, whereas the mouse devotes only 5% of its cycle to this stage. Overall duration is quite well related to nuclear DNA content – the more DNA the longer meiosis takes; but curiously enough, polyploidy leads to a shortening of the meiotic cycle, even though it increases the nuclear DNA content. Polyploid species have shorter meioses than related diploids. In the natural allopolyploid wheat series, meiosis lasts 42 hours in diploids, 30 hours in tetraploids and 24 hours in hexaploids. Moreover, a hexaploid wheat nucleus that has the same overall DNA content as a diploid *Allium* nucleus accomplishes meiosis in 24 hours, while the latter takes 96 hours. One is tempted to conclude that there is a simple positive correlation between nuclear DNA content and meiotic duration for a species at each polyploid level. However, such a conclusion would be of limited value, as we soon realize when we examine the results of experiments with mixed genomes of various forms of wheat and rye. These experiments do not uphold any simple relationship between ploidy and meiotic duration but they do indicate that specific genomes, chromosome sets, proceed through meiosis at quite distinctive rates and, when different genomes are mixed with one another, some can impose their pattern of timing on other genomes with which they may be artificially combined.

MEIOSIS IN REARRANGED CHROMOSOMES

The final matter that I wish to consider is meiosis and structural rearrangements. I have chosen to deal with this last because it provides excellent opportunities for students to test their understanding of the meiotic process. Someone who can confidently and correctly predict the consequences of structural rearrangements occupying different positions on the chromosomes and in combination with various distributions of chiasmata has truly come to terms with the meiotic process.

I shall give here only a few examples of the problem and then invite readers to imagine other circumstances and predict the consequences.

Chromosomal rearrangements may occur within single chromosomes or they may affect two or more members of the karyotype simultaneously. They depend on two events: breakage of chromosomes and reunion in a way that is different from the original one. We shall not consider how or when these breaks and reunions take place since that is a molecular matter and still very much the subject of varying levels of speculation. That is not to say it is unimportant. It has, in fact, been a matter of vigorous controversy for more than half a century. Nevertheless, our primary concern here is rearrangements and meiosis. Two kinds of rearrangements need to be considered: inversions and translocations.

Inversions are simply chromosome segments that have been turned around through 180°, so that their position in the sequence has been reversed. Single inversions are clearly two-break rearrangements if they involve an interstitial segment of the chromosome, but it is possible that they may be one-break events if they involve a segment from the breakpoint to the chromosome end. This, of course, begs the question of whether a chromosome end that normally shows no inclination to fuse with anything can suddenly change in such a way as to react and fuse with the freshly broken stump at some interstitial point on a chromosome arm.

If the breaks are on opposite sides of the centromere then the inversion is said to be **pericentric** (around and including the centromere) (Figure 3.22a). If both breaks are situated in the same chromosome arm then the inversion is a **paracentric** inversion (alongside the centromere but not including it) (Figure 3.22b).

It is important to realize that paracentric inversions remain invisible in mitotic chromosomes unless the chromosomes are stained in such a way as to show banding patterns or other kinds of linear differentiation. The paracentric inversion does not change the relative dimensions, arm ratio or centromere index of the chromosome. Pericentric inversions, on the other hand, do change the arm ratio of the chromosome, unless the two breaks are precisely equidistant from the centromere, so that they are capable of altering the karyotype of the organism and they are, for the most part, detectable in somatic cell divisions as well as in meiosis (Figure 3.23).

At meiosis, in an individual heterozygous for an inversion, complete pairing of homologous chromosomes will lead to a reversed loop at pachytene (Figure

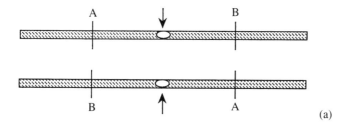

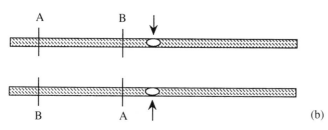

Figure 3.22 (a) A pericentric inversion involving the region A–B in a metacentric chromosome. Centromeres indicated by arrows. (b) A paracentric inversion involving the region A–B in a metacentric chromosome. Centromeres indicated by arrows.

3.24). If the inversion is pericentric, the centromeres will be inside the loop, if it is paracentric they will be outside. Loops usually only form in regions of quite large inversions. Small inversions usually simply prevent pairing over a short distance.

Translocations may involve two, three or four breaks and constitute the physical removal of a segment from one chromosome and its attachment to the

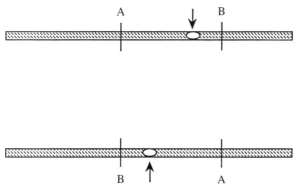

Figure 3.23 An asymmetric pericentric inversion involving the region A–B, changing a submetacentric chromosome to a metacentric one. Centromeres indicated by arrows.

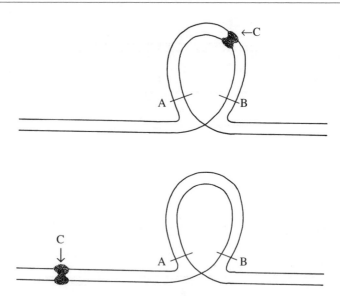

Figure 3.24 Homologous pairing of chromosomes heterozygous for inversions A–B by the formation of an inversion loop. (a) Pericentric inversion; (b) paracentric inversion. C, centromere.

end of another homologous or non-homologous chromosome or its reinsertion into a broken gap in another chromosome (Figure 3.25a). Reciprocal translocations involve actual exchange of segments between homologous or non-homologous chromosomes. Naturally, a reciprocal and equal translocation between two homologues will be of no consequence, beyond that which would follow from two cross-overs. An **unequal reciprocal** translocation involving homologous chromosomes will, on the other hand, generate two unequal-sized chromosomes from two that were originally the same (Figure 3.25b). Organisms that are heterozygous for translocations will show patterns of pairing at meiosis that reflect the propensity for individual segments of chromosomes to 'seek out', 'find', pair and form chiasmate associations with their homologous counterparts, no matter where they may have been translocated to in the karyotype.

Inversion and translocation heterozygosities really cause no problems, other than those that may arise on account of altered expression of relocated genes, unless chiasmata form within the rearranged segments. Without chiasmata, chromosome segments just pair up at zygotene and come apart again at diplotene. If chiasmata form within the rearranged segments, the situation is very different indeed. I shall deal here with just three examples, but readers are urged to experiment with karyotypes, inversions, translocations and chiasmate associations of their own inventions, so as to develop a clear understanding of the significance of these happenings in relation to the meiotic process.

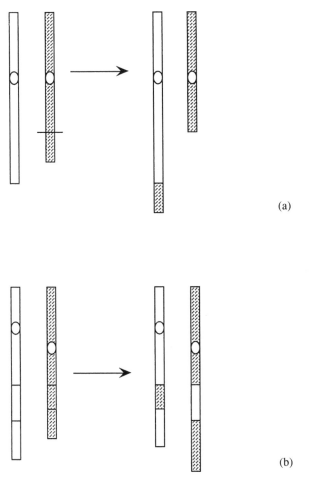

(a)

(b)

Figure 3.25 Two examples of translocations. In (a) the end of the long arm of the smaller chromosome has broken off and been translocated onto the end of the long arm of the larger one. In (b) we see a situation where two chromosomes have exchanged interstitial segments of unequal sizes, an event that is described as an unequal reciprocal translocation. These kind of events could involve homologous or non-homologous chromosomes.

Figure 3.26 illustrates the situation when a single cross-over occurs within a paracentric inversion. First, note that the diagram is of a chromosome with the centromere very near to one end; each homologue consists of two chromatids and these are closely paired, point for point, along their lengths. Next, identify the inverted segment, BC, and note that it is inverted on chromatids 3 and 4 belonging to homologue II. The inversion loop permits complete pairing from end to end of the chromosome arm. Carefully follow chromatids 1 and 3 from end to end and you will see how a normal chromatid (1) pairs with an inverted one (3).

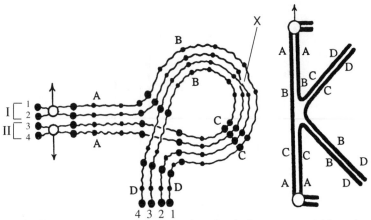

Figure 3.26 The consequences of the formation of a single cross-over (chiasma) forming within a paracentric inversion loop. *Trace each chromosome strand very carefully from end to end and note the formation of one chromosome with two centromeres and one without any centromere.* This leads to the formation at first meiotic anaphase of a **'bridge and fragment'**. (Reproduced with the kind permission of Cambridge University Press from White, M.J.D., *Animal Cytology and Evolution,* 1973.)

A cross-over has been introduced between chromatids 2 and 4 at the point marked X. Now carefully follow chromatid 4 from its centromeric end on the left of the diagram along to the right, through the cross-over and on to the other end – which is the centromeric end of chromatid 2. The paracentric inversion loop, together with the cross-over within the loop, has generated a chromatid with two centromeres, a **dicentric**. Now follow chromatid 2 from its terminus at the other end from the centromere up, into the loop, through the cross-over and back down to what was originally the end of chromatid 4. This time, we have a chromatid that lacks a centromere altogether and contains only the distal segments, DCBD, of the chromosome sequence. The result at first meiotic metaphase/anaphase will be a dicentric chromosome that will form an anaphase bridge and an acentric fragment. The bridge will break, most probably unequally. The fragment will be lost.

Next, examine Figure 3.27, which shows, on the same scheme, the situation that pertains when a single cross-over occurs in a pericentric inversion. Then consider what would happen if two chiasmata formed in a paracentric inversion, one involving chromatids 1 and 3 and the other chromatids 2 and 4. Then experiment with paracentric inversions having cross-overs both inside and outside the inverted segment and involving the same or different pairs of homologous chromatids.

This kind of exercise is of immense value as a means of firmly imprinting the basic rules of meiosis, the evolutionary consequences of structural rearrangements, the importance of chiasma localization in relation to rearrangements and, above all, the overall effectiveness of meiosis as a sieve or screening process

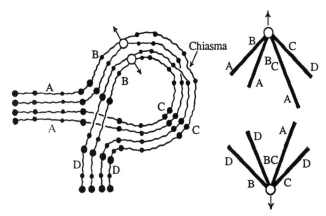

Figure 3.27 The consequences of a single cross-over (chiasma) forming in a pericentric inversion loop. *Trace each chromosome strand very carefully from end to end and note the formation of chromatids with unbalanced genetic constitutions.* Indeed, only one of the four chromatids retains all the normal factors in the right order, ABCD. (Reproduced with the kind permission of Cambridge University Press from White, M.J.D., *Animal Cytology and Evolution*, 1973.)

through which all chromosomes must pass before they and their genes are transmitted to the next generation by way of viable eggs and spermatozoa. The exercise can only be properly and profitably carried out with lots of paper and four coloured pencils and, thus equipped, it can be a stimulating and entertaining learning experience.

The situation with regard to translocation heterozygotes is much more complex but of immense evolutionary significance, since translocations are, if anything, more common than inversions. Suffice to say in the context of this introductory text that the consequences of translocations and associated cross-overs depend on a range of factors including the size of the translocated segment and its position in relation to the centromere, the positions of cross-over and the sizes and arm ratios of the chromosomes involved. The subject has been well covered by White (1973), to which the reader is specifically referred for further study of this topic.

RECOMMENDED FURTHER READING

Brinkley, B.R. (1991) Chromosomes, kinetochores and the microtubule connection. *BioEssays* **13**, 675–681.

Carpenter, A.T.C. (1987) Gene conversion, recombination nodules and the initiation of meiotic synapsis. *BioEssays* **6**, 232–236.

Evans, W. and Dickinson, H.G. (Eds.) (1983) *Controlling Events in Meiosis*, 38th Symposium of the Society for Experimental Biology. The Company of Biologists, Cambridge.

Henderson, S.A. and Koch, C.A. (1970) Co-orientation stability by physical tension: a demonstration with experimentally interlocked bivalents. *Chromosoma (Berl.)* **29**, 207–216.

Janssens, F.A. (1909) Spermatogenese dans les batraciens. V. La theorie de la chiasmatypie, nouvelle interpretation de cineses de maturation. *Cellule* **25**, 387–411.

John, B. (1990) *Meiosis*. The University Press, Cambridge.

John, B. and Lewis, K.R. (1986) *The Meiotic Mechanism*. Carolina Biological Supply Company, Burlington, NC.

Kezer, J., Sessions, S.K. and Leon, P. (1989) The meiotic structure and behavior of the strongly heteromorphic XY sex chromosomes of neotropical plethodontid salamanders of the genus *Oedipina*. *Chromosoma (Berl.)* **98**, 433–442.

Loidl, J. (1990) The initiation of meiotic chromosome pairing: the cytological view. *Genome* **33**, 759–778.

Nicklas, R.B. (1988) Chance encounters and precision in mitosis. *J. Cell Science* **89**, 283–285.

Rattner, J.B. (1991) The structure of the mammalian centromere. *BioEssays* **13**, 51–56.

White, M.J.D. (1973) *Animal Cytology and Evolution* . The University Press, Cambridge.

Sex chromosomes and sex determination

4

The great majority of higher animals have separate sexes and possess genetic sex-determining mechanisms. On the other hand, it is important to remember that there are several large groups of animals in which hermaphroditism is the rule and bisexuality the exception. Flatworms, earthworms, leeches and some molluscs are hermaphrodite, although within the same groups there are as many or more bisexual types. Hermaphroditism should in all cases be regarded as evolutionarily more advanced than bisexuality, and where we have a mixture of bisexual and hermaphrodite species within the same phylum or class, it is reasonable to suppose that the common ancestor was bisexual. Alternatively, we should consider carefully whether some of the present-day bisexual forms that have hermaphrodite relatives are descended from hermaphrodite ancestors and have reverted to bisexuality with redevelopment of a genetic sex-determining mechanism.

In this chapter we shall be concerned only with simple cytogenetic sex-determining mechanisms that involve one pair of chromosomes, most commonly referred to as an X and a Y. There are other sex-determining systems, some involving multiple sex chromosomes, some based on an interaction between sex-determining genes and environmental factors and some in which genetic factors have little or no role. Xs and Ys are simple and straightforward but we would be deluding ourselves if we regarded them as anything more than a convenient anchor point for studies of an impressive range of strategies that have evolved in connection with the determination of sex. Exploration beyond the level of XY systems might usefully commence with Michael White's *Animal Cytology and Evolution,* which represents one of the more compact yet thorough treatments of this complex field.

Where the genetic mechanism of sex determination resides in a single pair of chromosomes, these are homologous throughout their lengths in one sex (the X chromosomes) and non-homologous or only partly so in the other sex (X and Y).

The X is present in both sexes, the Y only in one. In some groups of animals the Y is absent altogether, notably in some of the locusts and grasshoppers, such that the diploid chromosome number in one sex is an uneven one.

The XX sex is said to be the **homogametic** sex since it produces only one kind of gamete while the XY or XO sex is **heterogametic** because it produces two different kinds of gametes in equal numbers. The letters X and Y refer to the sex chromosomes of animal species in which the male is the heterogametic sex. Where the female is the heterogametic sex, as in butterflies and moths, birds, some of the fishes, amphibians and reptiles, as well as representatives of several other invertebrate groups, the sex chromosomes are called **Z** and **W**, with the male being ZZ and the female ZW. Chromosomes that are not heteromorphic and not specifically connected with sex determination are commonly referred to as **autosomes**, thus the human diploid karyotype consists of 44 autosomes (22 pairs), an X and a Y chromosome.

There are two distinctly different types of XY sex-determining mechanisms. In the first, the sexual phenotype depends on the balance of female-determining genes on the X chromosomes and male-determining genes on the autosomes. *Drosophila* is the prime example. The Y plays no part in determining the sex phenotype, so that XO individuals are males and XX individuals with one or more Ys are phenotypically normal and fertile females. *Drosophila* with three sets of autosomes and just two X chromosomes are phenotypic intersexes, the extra set of autosomes having shifted the balance in the male direction. The same applies to other insects that are naturally XO/XX, such as locusts and grasshoppers. It does not apply to the examples of somatically XO mammals that we shall consider later in Chapter 6.

The other system of sex determination involves a strongly male-determining Y chromosome. *Homo sapiens* is as good an example as any. XO people are sterile females with ovarian dysgenesis or **Turner's syndrome**. XYYs are normal males, but may be unusually tall, aggressive and mentally subnormal. Individuals that are XXY, XXXY, XXYY and XXXXY are all basically male but all sterile, more or less abnormal and belong to a general type that is referred to as **Klinefelter's syndrome**.

X and Y or Z and W chromosomes are commonly, but by no means always, of completely different shapes and sizes. In *Homo sapiens* the X and Y are quite different. In chickens (*Gallus domesticus*), we have female heterogamety with the Z and W chromosomes again being very different in size. Amongst the tailed amphibians (Urodela), newts and salamanders, the large European newt *Pleurodeles waltlii* and the axolotl (*Ambystoma mexicanum*) have the female as the heterogametic sex and there are no morphologically distinguishable sex chromosomes. In the European newts of the genus *Triturus*, on the other hand, the male is the heterogametic sex and X and Y are distinguishable only in preparations where the chromosomes have been stained with the Giemsa C-banding technique, after which one of the pair of fourth longest chromosomes appears with a dark band near the end of its long arm (Figure 4.1). The band is

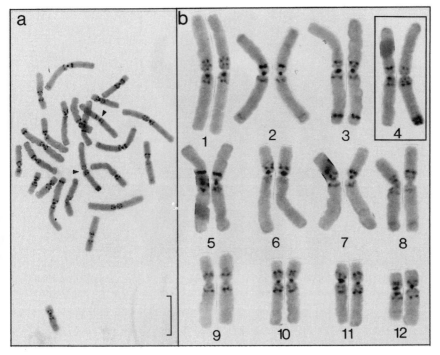

Figure 4.1 Giemsa C-banded metaphase of a spermatogonial mitosis in the newt *Triturus alpestris* (a) and the chromosomes of that same metaphase arranged in a karyotype (b). Only one of the two 4th chromosomes (arrows in a) has a darkly staining band of heterochromatin at the tip of its long arm. Otherwise the two 4th chromosomes are identical. This Y-chromosome-associated, Giemsa C-positive heterochromatin is considered to represent an early stage in the evolution of sex chromosome differentiation. The scale bar represents 10 μm.

absent or much smaller in the corresponding position of the partner chromosome. Otherwise, chromosomes X and Y are morphologically indistinguishable.

Amongst the lungless salamanders of North and Central America (family Plethodontidae), some species have no morphologically identifiable sex chromosomes while others have distinct heteromorphisms. The Central American genera *Oedipina* and *Thorius,* for example, have sex chromosomes that are immediately identifiable and especially striking in appearance (Figure 4.2), partly, of course, because they are amongst the largest chromosomes in the world, coming from animals that have enormous genomes (more than 20 times as large as the human genome) and relatively low haploid chromosome numbers (13 or 14). The X and Y chromosomes in plethodontids have a short common segment near to one end, and this is the only region where pairing takes place between X and Y during meiotic prophase (Figure 4.3).

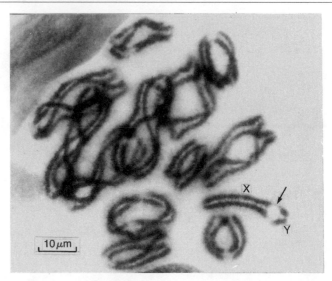

Figure 4.2 Diplotene of meiosis 1 in the male salamander *Oedipina uniformis*. X and Y chromosomes are shown in end-to-end synaptic contact (arrow). The strikingly different shapes and sizes – heteromorphism – of the X and Y is shown in these tropical salamanders from Central America better than in any other species of amphibian. (Reproduced with the kind permission of Dr James Kezer, University of Oregon, and Springer-Verlag.)

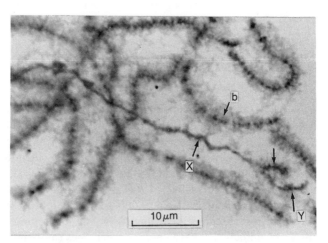

Figure 4.3 Part of a spread of pachytene chromosomes from *Oedipina uniformis* male, showing the fully synapsed pachytene bivalents (b) contrasted with the long single X chromosome synapsed only at its end (unlabelled arrow) with the short Y. (Reproduced with the kind permission of Dr James Kezer of the University of Oregon, and Springer-Verlag.)

The behaviour of X and Y chromosomes is quite well typified in mammals (Figure 4.4). In humans and mice, for example, a common segment is located right at the ends of the short arms of both X and Y. Meiotic pairing and genetic recombination regularly occur within this region, but only exceptionally – and abnormally – in other regions of the chromosomes. It is particularly interesting to note that in about 1/20 000 live human births we see the effects of recombination between X and Y outside the common region. The resulting phenotypic abnormalities, XX males and XY females, not only prove the male-determining force of the Y chromosome but tell us something about where that force resides in the Y. The XY females have lost the male-determining region of the Y, which has been transferred by genetic crossing over to the X that is responsible for the XX males.

At meiosis 1, X and Y and Z and W separate from one another and go to opposite poles of the spindle. If they fail to disjoin in this manner then XX, XY and 'OO' gametes will be generated which, in humans at any rate, will be potential sources of individuals with XXY, XYY and XO constitutions.

In those groups where many species have XO males, it is most likely that

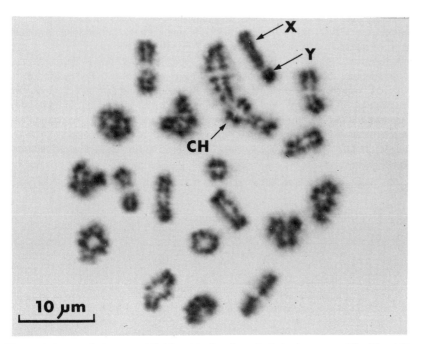

Figure 4.4 Late diplotene or 'diakinesis' of male meiosis in the mouse. The X and Y chromosomes are seen in end-to-end contact at the top right of the picture. This particular animal possessed a translocation (T-27), which presents here as a chain (CH) of four chromosomes (two pairs). (Picture kindly provided by C.V. Beechy, G. Breckon and A.H. Cawood of the Medical Research Council Radiobiology Unit at Harwell.)

ancestral types possessed an XY system and in the course of time the Y was of diminishing importance and was eventually lost altogether. Sometimes, however, it comes back again, which provides us with some interesting material for an elementary lesson in the evolution of sex chromosomes.

The usual way in which such transformations have come about is through centric fusions between the acrocentric X of an XX/XO form and an acrocentric autosome. A fusion of this kind will produce a 'neo-X' chromosome. When the fusion has become fixed in the population and no more unfused Xs exist in the population, the original acrocentric autosome – the partner of the autosome that is now irrevocably joined to the original X – will be confined to the male line (or the female line in groups with female heterogamety) and it will constitute a 'neo-Y'. Of course, it will not, at least at first, have any sex-determining genes on it.

RECOMMENDED FURTHER READING

Goodfellow, P.N., Craig, I.W., Smith, J.C. and Wolfe, J. (Eds.) (1987) *The Mammalian Y Chromosome: Molecular Search for the Sex Determining Factor. Development,* Vol. 101 Supplement. The Company of Biologists, Cambridge.

Marshall Graves, J.A. and Watson, J.M. (1991) Mammalian sex chromosomes: evolution of organization and function. *Chromosoma (Berl.)* **101**, 63–68.

Marshall Graves, J.A. (1987) The evolution of mammalian sex chromosomes and dosage compensation – clues from marsupials and monotremes. *Trends Genet.* **3**, 252–256.

Schmid, M., Nanda, I., Steinlein, C., Kausch, K., Haaf, T. and Epplen J.T. (1991) Sex determining mechanisms and sex chromosomes in amphibia. In *Amphibian Cytogenetics and Evolution*. D.M. Green and S.K. Sessions (Eds.). Academic Press, New York and London, pp. 393–430.

White, M.J.D. (1973) *Animal Cytology and Evolution*. The University Press, Cambridge, Chapters 16–20.

Polyploidy in animals

5

Being polyploid means having more than the normal diploid number of chromosomes in all cells of the body, including the cells that will pass through meiosis in the gonads and form gametes.

There is no doubt whatever that a substantial number of animal species have become polyploid in the course of their evolution. However, when we look at the kinds of animals that are involved and consider their biology, two rules are plain to see. First, many polyploids are either hermaphrodite or **parthenogenetic** (Greek: *parthenos*, virgin; *genesis*, descent). Polyploidy is rare amongst bisexual species. Secondly, where polyploidy has become established in a bisexual species, then it has been closely followed or accompanied by certain adjustments in the meiotic system such as to ensure that the increase in the number of sets of chromosomes does not prejudice the chances of these chromosomes segregating or disjoining in a balanced manner. Our biggest problem, when we have examined some examples of polyploidy and its associated biological and molecular features, will be that of deciding the order in which it has evolved in relation to parthenogenesis or meiotic adjustment. In essence, animal polyploidy presents us with a remarkably good opportunity to test our understanding of meiosis and the selective forces that lie behind the evolution of reproductive strategies in animals.

THE PROBLEMS OF BECOMING POLYPLOID

That polyploidy is rare amongst bisexual species suggests that it may be incompatible with the sexual process. To a large extent this is true. For example, there are bound to be difficulties facing polyploid organisms with obligatory cross-fertilization. If, as a consequence of a meiotic or developmental accident, a single animal becomes **tetraploid** (**four** sets of chromosomes per cell), then it will produce only diploid gametes, and in conjunction with its normal diploid mates in the same population, all of which of course have haploid gametes, it will be capable only of producing **triploid** (**three** sets of chromosomes) progeny, $2N + 1N = 3N$; and this provided the tetraploid manages to accomplish a balanced meiotic disjunction of its two diploid sets of chromosomes.

The triploid progeny will be incapable of producing haploid or diploid gametes, since the chances of a balanced meiotic disjunction of three sets of chromosomes, into one full haploid set and one full diploid set, are remote. These points would seem utterly to preclude the establishment of tetraploidy in a bisexual species. One tetraploid can make no impression on its own, and the chances of two arising in the same population within the same generation and mating with one another are definitely not good. Yet we shall soon consider good evidence for the establishment of tetraploidy in several species of bisexually reproducing animals. How did it happen?

Perhaps the most obvious factor that operates against polyploidy in evolution of some species is the chromosomal sex-determining mechanism. Where sex is determined by genes that reside on specific heteromorphic sex chromosomes, polyploidy will abolish the heterogamety upon which sex determination depends. Consider carefully what will happen if an XX/XY species becomes tetraploid. If it were the male that became polyploid then it would be XXYY. The two Xs and the two Ys would form respectable bivalents at meiosis and segregate in a balanced reductional manner to give gametes that were all XY. So at one swift step we have lost the chromosome heteromorphism upon which sex determination depends. If our tetraploid mates with a diploid female, then all the progeny will be XXY. In the unlikely event of its being able to mate with a tetraploid female, then all the progeny will be XXXY.

A third factor that would seem to discourage polyploidy is the need for balanced disjunction of chromosomes at meiotic anaphase 1. The difficulties will immediately be apparent with regard to uneven polyploids such as triploids. If trivalents form at first meiosis, then the daughter cells will inevitably receive unbalanced sets of chromosomes; there will be two of some chromosomes and just one of others in each cell and the entire situation will be altogether intolerable. If trivalents do not form, then there will be univalents, and we have already seen what happens to them at first anaphase (Chapter 3) (Figure 5.1).

In the case of even polyploids, such as tetraploids, all will be well provided some mechanism or circumstance exists to encourage the formation of bivalents at meiosis, or the formation of quadrivalents of a kind that are able to disjoin such that two chromosomes go to one pole and two to the other at first meiotic anaphase. That will happen, and tetravalents will behave themselves, just so long as any one chromosome is not involved in chiasmate association with more than two others in the set of four; which is to say that tetraploidy is more likely to pass 'the meiotic test' in organisms that have small chromosomes and an average chiasma frequency of less than two per chromosome.

How, then, can a species become polyploid and stay alive? It can overcome the bisexual problem by being hermaphrodite or parthenogenetic, or by reducing to one the number of decidedly male- or female-determining chromosomes in its polyploid chromosome set, and thereby re-establishing a sex chromosome

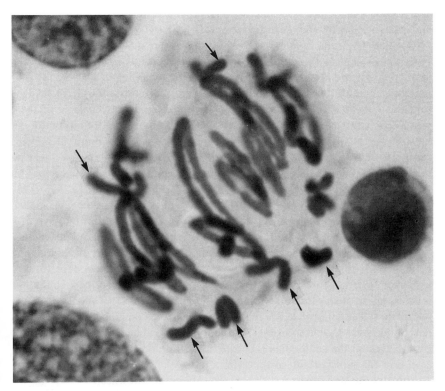

Figure 5.1 A first meiotic metaphase in side view from a male triploid hybrid newt *Triturus marmoratus* × *T. karelini*, photographed from a preparation made by L.A. Lantz and H.G. Callan in 1954. The metaphase shows 12 bivalents and a further 12 univalents that will distribute randomly between the daughter cells, most probably giving rise to secondary spermatocytes that have unbalanced chromosome complements.

heteromorphism. It could overcome the meiotic problems by discouraging the formation of uneven multivalents or by encouraging the effective 'diploid-ization (bivalents only)' of meiosis: easy and hypothetical words outlining events that must surely represent massive and extraordinary evolutionary happenings. How *do* the first polyploids in a population establish and propagate themselves? How *does* an organism become parthenogenetic? How *is* meiotic pairing and chiasma formation between chromosomes regulated? With these sorts of matters in mind, let us now look at some specific examples.

SOME SPECIFIC EXAMPLES

Our first example is a frog belonging to the genus *Odontophrynus*. The characteristic haploid number for this genus is 11. There are plainly diploid species, such as *O. cultripes* and *O. occidentalis*, where everything is normal and straightforward, and there are tetraploid forms, like *O. americanus*, where 2N=44 and where the amount of DNA per somatic cell nucleus is approximately twice that found in the diploid species. *O. americanus* is presumed to be a tetraploid. Its chromosomes fall into 11 groups of homologues, each having four members, and at meiosis up to 11 quadrivalents may form. Furthermore, it is said that *within the same species* there are normal diploid individuals. Here, then, we have a bisexual animal that has succeeded in becoming an **autopolyploid**, but has not yet fully adjusted its meiotic mechanism in so far as multivalents still form. However, the risks associated with multivalents in this case must be minimal. The animal has a small genome (1.5 pg) and an average chiasma frequency of less than two per chromosome.

Our next example comes from the family of fishes known as the Cyprinidae. The carp (*Cyprinus carpio*) and the goldfish (*Carassium auratus*) have 104 chromosomes per somatic cell, whereas other members of the family have only 44–54. The amounts of DNA per nucleus are correspondingly different. The conclusion is that carp and goldfish became polyploid at some stage in their evolution. During meiosis in these fish, no multivalents are formed, so that we must suppose that a meiotic diploidization has been superimposed on the tetraploid system. In plants, where polyploidy is more common in the absence of heteromorphic sex chromosomes and amongst the abundance of monoecious (both sexes on the same plant) species, it is said that there has been a strong selection for **multivalent suppressor alleles**. We have already seen a good example of this in Chapter 3 with regard to allohexaploid wheat.

A good example of thoroughly established polyploidy can be found in the genus *Xenopus*, the so-called African clawed toads. The genus includes 15 species and five subspecies and chromosome numbers in somatic cells range from 20 to 108, with corresponding differences in C-values. Some examples are shown in the following table.

Species	Chromosomes per somatic cell	DNA per nucleus (pg)
X. tropicalis	20	3.5
X. epitropicalis	40	6.0
X. laevis	36	6.4
X. vestitus	72	12.8
X. ruenzoriensis	108	16.3

Xenopus polyploids are almost certainly **allopolyploids**, meaning that they have arisen by joining together of the chromosome sets of two or more species, in the same manner as we have already seen in the allohexaploid variety of *Triticum* (common bread wheat). In all of them, meiosis is normal and fully 'diploidized' in the sense that no multivalents form. Conclusive proof of polyploidy in *Xenopus* and a satisfactory explanation of the crucial 36/40 difference between *laevis* and *epitropicalis* will soon be forthcoming from studies of high-resolution Giemsa G-banding of the chromosomes in these species.

There are, of course, many other examples of suspected polyploidy amongst the vertebrates, most of them based on suggestive differences in chromosome number, amounts of DNA per nucleus, numbers of comparable gene loci as determined by numbers of electrophoretic variants of certain proteins and the occurrence of multivalents during meiosis. None of these criteria are entirely dependable. Differences in chromosome number, for example, may only reflect a wide range within a family. It used to be said that the golden hamster (*Microcricetus auratus*, N=22) was a polyploid derived from two other species in which N=11. Yet there is a whole series of cricetine rodents with chromosome numbers ranging from 11 to 22. Amounts of DNA can be just as misleading: C-values can vary widely from species to species within a particular genus.

EVOLVING POLYPLOIDY

One of the most remarkable cases of evolving polyploidy is found in the common European freshwater flatworm, *Dugesia lugubris*. This animal is hermaphrodite, with both ovary and testis forming in the same individual but, since self-fertilization does not occur, it might be said to be effectively bisexual. The species comprises seven **biotypes** that are distinguishable from one another on the basis of their chromosome numbers in somatic and germ cells and their meiotic or ameiotic behaviour. The first four biotypes are particularly interesting in the present context. Their constitutions are as follows:

The basic chromosome complement of N=4 (haploid) is the same in all four biotypes. A is normal in every respect. B is a **somatic triploid** with hexaploid

Biotype	Somatic cell chromosome number	Germ cell chromosome number	Meiosis
A	8	8	Normal in both ovary and testis
B	12	24 ovary 8 testis	Normal in both ovary and testis
C	12	12 ovary	Non-reductional in ovary; normal in testis
D	16	16 ovary 8 testis	non-reductional in ovary; normal in testis

oocytes that pass through a perfectly respectable meiosis to revert to triploid eggs, which then develop without the addition of a set of chromosomes from the male parent – a phenomenon known as **gynogenesis**. C is a somatic triploid that has shortcut the meiotic process by confining its oocytes to a single non-reductional maturation division, and once again, it is gynogenetic. D is a tetraploid that has done the same thing.

The most astonishing feature of the story is the conservatism of the male gametogenic system. Its apparently essential diploidy is faithfully maintained in all four biotypes by elimination of one (in the triploids) or two (in the tetraploids) sets of chromosomes from the polyploid, premeiotic, primordial germ cells that will go to make up the testis.

Biotypes A, B, C and D can all interbreed with one another, although it is only fair to point out that the sperm is only genetically significant in biotype A. All the others are **pseudogamic**, in the sense that the sperm merely activates the development of the egg.

Polyploidy allied to gynogenesis and pseudogamic reproduction is also common in earthworms, although not those of the genus *Lumbricus*, and it has been described in several species of amphibians and reptiles. Jefferson's salamander, *Ambystoma jeffersonianum*, named after the third – the naturalist – president of the United States, Thomas Jefferson (1743–1826), has diploid males and females that are normal in every respect. They have a haploid chromosome number of 14. Living amongst the diploid individuals there are also triploid females in which the last oogonial mitosis is endomitotic (unaccompanied by nuclear or cell division), so producing hexaploid oocytes, each of which has a complement of 84 chromosomes. It is worth noting here that throughout oogenesis all of these chromosomes are of the **lampbrush** type that is typical of the growing oocytes of amphibians, reptiles and birds as well as many types of invertebrate. We shall be looking closely at lampbrush chromosomes in Chapter 9. The chromosomes form only bivalents, which says that the animal has effectively 'diploidized' its meiotic mechanism. It is not yet clear whether these triploid females are allo- or autopolyploids. The meiotic bivalents show a high chiasma frequency, but since the chiasmata probably join identical chromosomes that are direct products of the preceding oogonial endomitosis, their importance is likely to be restricted to ensuring proper segregation and disjunction at metaphase 1.

These, then, are but a few examples of the kinds of polyploidy that we encounter amongst animals. They present us with two major questions. One is the matter of which came first, polyploidy or parthenogenesis; unanswerable but nonetheless debatable. The other relates to chromosome pairing, for all that has to be done to get a polyploid through the meiotic test is to eliminate or drastically reduce the tendency for multivalents to form during meiotic prophase, a step that is particularly important in an animal like *Ambystoma* in which the chromosomes are large and chiasmata are frequent.

RECOMMENDED FURTHER READING

Ohno, S. (1970) *Evolution by Gene Duplication.* Springer-Verlag, Berlin, Heidelberg, New York, Chapter 16, pp. 98–105.

White, M.J.D. (1978) *Modes of Speciation.* W.H. Freeman, San Francisco, Chapter 8, pp. 261–285.

Inactivation and elimination of chromosomes | 6

In this chapter we shall be dealing specifically with a phenomenon that is sometimes referred to as **facultative heterochromatin**. The term is generally applied to a minority component of chromosomal material (chromatin) that looks different (hetero-) from the remainder on account of being condensed to a greater or lesser degree in certain cell types. The word facultative means that the differential condensation is not evident in all cell types and the regions that are heterochromatic may, in some circumstances, appear no different from the remainder of the chromosomal material. The other type of heterochromatin, referred to as **constitutive**, is regarded as a permanent feature of the organism's karyotype based on some molecular and/or structural characteristic of the chromosomal nucleoprotein in the heterochromatic regions. Giemsa C-bands, for example, are designated constitutive heterochromatin because they are based on repeat-sequence DNA that is part of the organism's permanent, heritable genetic constitution. We shall be looking more closely at some of the properties of constitutive heterochromatin in Chapter 10.

THE BARR BODY

Facultative heterochromatin is not a newly discovered phenomenon, but it was regarded with little more than academic interest until 1949, when Barr and Bertram identified a 'sex chromatin body' in nuclei of the tissues of female cats. They said it represented the fused heterochromatic segments of the two X chromosomes of the female mammal. They were, of course, quite wrong, but their observations nevertheless stimulated interest in mammalian sex chromatin and led other people to look more carefully at sex chromosomes in mammalian systems. A little later, a single X chromosome was shown to stain differentially

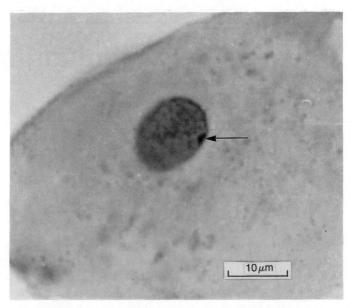

Figure 6.1 A simple preparation made in a few minutes by scraping the inner surface of the cheek with a wooden spatula, smearing the scraped material onto a clean microscope slide and immediately flooding it with a solution of orcein in 45% acetic acid (see Appendix, page 206), covering with a clean coverslip and examining with bright-field microscopy. In such a preparation, the smear is dominated by large oral mucosal cells with their round granular nuclei. If the preparation is made from a normal female person, each nucleus shows a compact half-moon-shaped Barr body closely adhering to the inner surface of the nuclear envelope (arrow). This represents the inactivated second X chromosome. *Compare the size of this Barr body with that of the X chromosome or the average C-group chromosome in Figures 2.3 and 2.4.*

in late prophase and early metaphase in diploid liver cells (hepatocytes) of rats. This was rightly supposed to correspond to the sex chromatin body of interphase nuclei. Later still, Herbert Taylor, of repute through his studies of semiconservative replication in *Vicia* (see Chapter 7), was able to show that the two X chromosomes of female hamsters replicated at different times in S-phase. The final stage in the opening part of this story came when Mary Lyon made the correct suggestion that the so-called Barr or sex chromatin body was an inactivated sex chromosome. She established that inactivation happens quite early in development, that it affects either of the two Xs at random, and that once it has happened to an X chromosome it is permanent in the sense that all the descendants of that chromosome throughout subsequent mitoses will be inactivated. Lyon also proposed that X inactivation was common to all mammals. Figure 6.1 shows the Barr bodies in a cell preparation from a normal female person.

X INACTIVATION IN MICE AND MULES

The experiments that confirmed the main features of the Lyon hypothesis are now regarded as classics. In mice, there are genes for coat colour on the X chromosome (**X-linked**). Mice that are heterozygous for these genes have variegated coats, explained by supposing that in some cells one X chromosome is active and a normal coat colour gene is expressed, whilst in other cells the other X is active and a mutant coat colour gene is expressed. Inactivation of the X chromosome in this instance must have taken place after the initial differentiation of the cells that will eventually be responsible for laying down the pattern of coat colour. If inactivation had happened before that point, then all cells would have had the same X inactivated and the coat would be uniformly of one colour.

Also in mice, the X chromosome can be 'tagged' by translocating a segment of an autosome onto it, so making it easily recognizable as the largest chromosome in the set. The autosomal segment carries two genes for coat colour, different from the ones on the X chromosome. Females heterozygous for the translocation and also for the wild-type and recessive alleles of the two autosomal coat colour genes have variegated coats, so proving that the translocated autosomal segment is forced to conform and suffer inactivation along with the X onto which it has been translocated. If this were not so and if the translocated autosomal segment were not affected by inactivation, then all cells would express the dominant wild-type alleles of the autosome-linked coat colour genes.

In the female mule derived from a female donkey (*Equus asinus*, 2N=62) and a male horse (*Equus caballus*, 2N=64), one X is a medium-sized metacentric derived from the horse and the other, derived from the donkey, is a large highly asymmetric chromosome with its centromere near to one end. The two Xs are therefore clearly distinguishable on account of their sizes and shapes, and there are few problems in showing that inactivation is random and that the inactivated chromosome replicates late in S-phase.

The process of X inactivation is believed to have evolved as a system of **dosage compensation**, such that although the two sexes have different numbers of X chromosomes there is only a single X *active* in adult somatic cells of both males and females. The compensatory mechanism is equally effective in individuals where there are more than two X chromosomes. The human triple X female, for example, has two inactivated X chromosomes and two Barr bodies are visible in each of her cells.

THE TIMING OF INACTIVATION

Two more points are important in relation to the basic 'rules' of inactivation. First, it is reversible. The Xs of cells in the germ line, spermatocytes and oocytes, are not inactivated. The Xs of some tissues in old people are not inactivated. Secondly, it is not entirely true to say that it is random. At an early stage in mammalian development the embryo takes the form of a trophoblast, which is a hollow ball of cells made up of trophectoderm and a small knot of cells, the embryo proper, associated with the inner surface of the trophectodermal wall (Figure 6.2).

X inactivation first happens in the mouse trophectoderm cells 4 days after fertilization. The chromosome that is inactivated in these cells is *always* the one that was contributed by the embryo's father, the **paternally derived** one. One or two days later, random X inactivation happens in the cells of the embryo. In humans, non-random paternal X inactivation first takes place in trophectoderm cells at around 13–14 days after fertilization, followed by random inactivation in the embryonic cells at around 18 days. Nature was never so simple as to confine herself to just one uniform strategy and, as we shall soon see, random X inactivation is to be regarded as an exception rather than a rule.

THE ROLE OF THE SECOND X

It is important to note that X inactivation and dosage compensation do not signify that the second X of the female mammal is dispensable, for it certainly is

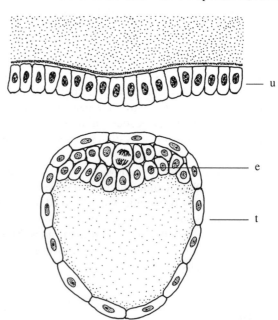

Figure 6.2 The trophoblast stage of a mammalian embryo showing the trophectoderm layer, where non-random X inactivation first takes place (t) and the inner cell mass which gives rise to the embryo (e), just before implantation into the uterine wall (u).

		Parents		
	Male XY		Female X	
Gametes	X	Y	X	O
Progeny	XX	XO	XY	YO
Prognosis	Female	Female	Male	Inviable
	Abnormal?	Normal	Normal	

not. Two quite simple observations tell us that it is very important indeed. First, in mammals, the Y chromosome is absolutely essential to maleness and a single X chromosome is essential for viability. Some simple juggling with Xs and Ys with a pencil and paper is all that will be needed to demonstrate that a species that starts out with XY males and XO females will soon produce a preponderance of females, half of which will have substantially reduced fertility (see table above). Such a species is hardly likely to be successful!

So the second X is essential, if only to act as a 'balancer' for the Y.

The significance of this matter comes clearly into perspective when one examines the lengths an organism has to go to in order to adopt a stable and successful XO female constitution. *Microtus oregoni* (the creeping vole) is a good example of one of the very few well-established XY/XO rodents. In this little mammal the male begins life with XY (2N=18) and all his somatic cells remain that way. However, in the primordial germ cells of the fetal testis there is a preferential non-disjunction of the X to produce cells that are either XXY or OY. We should be absolutely clear that this non-disjunction event takes place during the spermatogonial mitoses and not during meiosis. Subsequently, only the OY cells differentiate to produce functional gametes, which will therefore have either an O or a Y constitution. The sperm that has no sex chromosome will be female determining. The one that has the Y will be male determining. But that is not the end of the story. The XO constitution of the female is maintained in all her somatic cells, but once again, in the primordial germ cells of the fetal ovary, there is a preferential non-disjunction of the X. Oogonia are produced that are either XX or OO, and of course only the former survive to make eggs that are uniformly X, thus avoiding a doubly heterogametic condition in which male and female both make two kinds of gametes. Undoubtedly, life is much simpler with a straightforward XX/XY system. Or is it? Which strategy is likely to be the more successful? What might have been the evolutionary pathways along which each strategy was developed? Is one of these strategies likely to be derived from the other or have both evolved independently? What *were* the selection pressures that led to X inactivation or elimination?

The other observation that confirms the importance of the second X chromosome, at least in humans, is Turner's syndrome, a quite severely abnormal phenotype that is associated with an XO constitution. It is worth noting, however, that XO mice are apparently normal and fertile, and doubtless the same applies to many other mammals that remain to be investigated.

THE GENETIC AND MOLECULAR BASIS OF INACTIVATION

The mechanism of X inactivation remains unclear although we know enough about it now to be able to ask sensible questions and identify good targets for future investigations.

How do active and inactive X chromosomes differ? They differ in the specific pattern and extent of **methylation** of their DNA. Methylation means the substitution of 5-methylcytosine for cytosine during DNA replication in places where the sequence 5'-*a*CG*b* occurs, where *a* and *b* can be any of the four nucleotides. Not all the CGs are methylated. The pattern of methylation, once established, is maintained faithfully for generation after generation of cells. A high level of methylation is generally a property of inactive genes. The two Xs also differ in chromatin structure such that the inactive X replicates late and looks different from the active one: one is condensed and the other is not.

Are there any genes involved in X inactivation? Yes indeed there are. The process of inactivation seems to be controlled by a locus called the '**X inactivation centre (*Xic*)**'. Pieces of the X, either alone or translocated onto autosomes, do not undergo inactivation if they do not include the *Xic* locus. The *Xic* locus includes a gene, *XIST*, that has the unusual property of being expressed only from the inactivated X. *XIST* was found during a search for genes that escape inactivation on the human X chromosome. Unlike all other genes that were examined, it turned out not to be expressed in normal human males or in XO females, but it was expressed in normal females and in patients with multiple X chromosomes. Moreover, unlike a number of other genes that are expressed from the inactivated X, *XIST* has no homologous counterpart on the Y chromosome.

INACTIVATION AND IMPRINTING

Mary Lyon and her associates and followers showed that in some common mammals, including man, X inactivation and late replication in the female are events that affect the Xs of paternal and maternal origin at random. We do not know that this condition prevails in all eutherian mammals that have an XX/XY constitution. Only painstaking investigation over an even wider range of species will tell. What we do know is that the Metatheria have done things a little differently, and it is important to examine their strategies before we attempt to sort out this whole rather complex range of phenomena that involve sex chromosomes, inactivation and dosage compensation.

Euros (*Macropus robustus exubescens*, 2N=16), wallaroos (*M. r. robustus*, 2N=20) and red kangaroos (*Megaleia rufa*, 2N=20) can all be crossed to yield hybrid animals and, just as with horses and donkeys, the X chromosomes of each species are easily distinguishable on the basis of size alone. Three different crosses have been accomplished, and the replicative chronologies of the X chromosomes have been examined along with characteristics of the enzyme glucose-6-phosphate dehydrogenase (G6PD) which is X-linked and species specific in its electrophoretic mobility. The results were clear. Whichever way the crosses were made, it was always the paternally inherited X chromosome that was late replicating and inactive in the sense that it failed to make G6PD.

Two other species of marsupial show an even more uncompromising situation in which an X chromosome is not just inactivated in the female, but totally eliminated from the somatic cells. In the two animals concerned, bandicoots belonging to the genera *Parameles* and *Isoodon*, all somatic tissues show 2N=13 and an XO constitution, whereas all mitotic figures from the testis and ovary show 2N=14 and an XX or XY constitution respectively. In these species it seems that the male starts out with XY and eliminates the Y from all cells except those that will go to form the testis. The female starts out with XX and eliminates an X from all cells except those that will go to make up the ovary. Unfortunately it is not known whether elimination is random or paternally biased, but it seems extremely likely that it will be the latter. What is particularly significant about this situation is that both X and Y chromosomes are involved and it represents a decisive step from inactivation to elimination.

The important difference between X inactivation in eutherian mammals and X inactivation or elimination in Metatheria is that the former is a random process that seems to have evolved principally as a dosage compensation mechanism. The latter reflects an inherent property of a paternal genome that is absent or different in a maternal one. Differential behaviour of all or a part of the paternal genome signifies that there has been a specific **imprinting** of the paternal genome, most probably during spermatogenesis, such that the maternal and paternal genomes from which an embryo is formed are not functionally equivalent. In this context it is perhaps interesting to note that mammalian embryos that possess either two paternal genomes or two maternal ones fail to develop. The presence of both a male and female nucleus is essential in an egg for full-term development.

Imprinting, however, is not really a mammalian invention. It was certainly around amongst the insects long before mammals ever came on the scene. To see differential chromosome behaviour, based on imprinting, on a truly spectacular scale we must look to the family of fungus gnats, the Sciaridae, and more specifically to the species *Sciara coprophila*. Let us begin at the beginning with the zygote. Here there are three pairs of autosomes, three large metacentric 'limited (L)' chromosomes, and three X chromosomes (Figure 6.3).

Two of the Xs came from the father and one from the mother. Four ordinary cleavage mitoses take place, making 16 cells, with all the chromosomes behaving perfectly normally. Then at the fifth or sixth cleavage mitosis all the L chromosomes are eliminated from all cells except those few that will eventually give rise to the ovary or testis: the eliminated chromosomes simply attach to the spindle, fail to split into their two chromatids, are left behind at anaphase, excluded from the daughter nuclei and disappear. Then at a later division, one paternal X is eliminated in animals that are to become females and both paternal Xs are eliminated in animals that are to become males. This latter step might be regarded as the essential one for sex determination, since it leads to male and female being both karyotypically and phenotypically different from one another. Here we have differential elimination of paternal Xs just as we have seen in the metatherian mammals, but for the purpose of sex determination rather than dosage compensation.

In the cells that are to form the eggs and sperm of *Sciara,* events are unbelievably complex and contrived. In both the male and the female we have elimination of one or two of the L chromosomes followed by elimination of one of the paternal Xs. Therefore, oogonia and spermatogonia have precisely the same complement of chromosomes. Female meiosis is normal and entirely respectable, so that eggs will have one of each autosome, one or two L chromosomes and one X chromosome. It is during spermatogenesis that the fun begins! Remember that the spermatogonium starts out with three pairs of autosomes, one or two L chromosomes and two X chromosomes, one of which is of maternal origin. The first meiotic division, if indeed it can be called that, takes place on a cone-shaped half-spindle. It is reductional in the sense that it eliminates all the remaining paternally derived chromosomes; that is an X, the paternally derived autosomes and any paternally derived L chromosomes. The second meiotic division is normal and equational for all chromosomes except the X, both chromatids of which pass to the same pole. The sperm therefore comes to contain three autosomes, one or two L chromosomes and two X chromosomes, all of which are maternally derived.

Sciara, then, presents us with the following phenomena:

1. differential condensation of L chromosomes;
2. differential elimination of L chromosomes in somatic-line cells;
3. selective elimination of paternal L chromosomes in germ-line mitosis;
4. preferential elimination of specific numbers of paternal X chromosomes in somatic-line mitosis;
5. loss of one paternal X chromosome in germ-line mitosis (this event is distinctive in so far as the chromosome is said to be pushed out through the nuclear envelope rather than being eliminated by misbehaviour on the mitotic spindle);
6. selective segregation of maternal from paternal chromosomes at male meiosis 1, but perfectly normal female meiosis 1;
7. non-disjunction of the two X chromatids at second male meiosis.

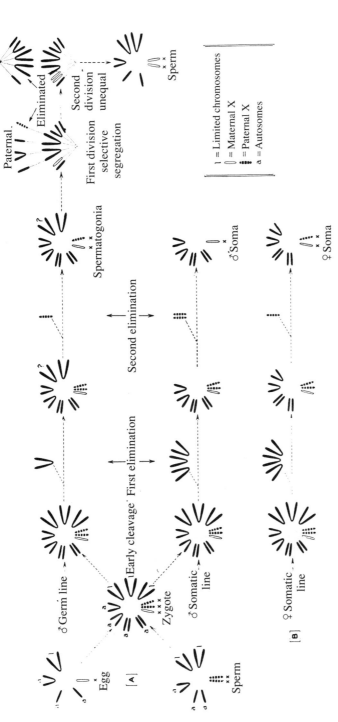

Figure 6.3 The chromosome cycle of *Sciara coprophila* during fertilization, cleavage and spermatogenesis. The maternal X chromosomes that go into the sperm become, of course, the paternal X chromosomes at fertilization. The number of 'limited' chromosomes actually varies from 1 to 3 amongst individuals and populations. Diagram A serves to represent equally well the happenings in the female, except with regard to gametogenesis, which is normal, and the second chromosome elimination from the somatic line, where only one and not two, of the X chromosomes is eliminated. (Reproduced from Metz, C.W., *Am. Nat.* **72**, 485–520, 1938.)

There is no point in attempting an explanation of all this, but at least let us see if we can classify the events and so perhaps bring certain manageable questions into focus. In 1 and 2 above we have differential condensation coupled with abnormal chromosome behaviour leading to eliminations in mitosis: these are not new phenomena. Later in this chapter we will examine some classical examples of this kind of event. In 3 we have an adaptation to ensure that the number of L chromosomes does not increase from generation to generation. Note that the total behaviour of the L chromosomes is different according to whether they are in a germ-line cell or a somatic-line cell. In 4 we have a curious system of sex determination, for it is this step that decides whether the organism will develop as a male or a female. In 5 we seem to have a unique strategy for getting rid of the paternal X in the male germ line. In 6 and 7 we have the idiosyncrasies of male meiosis, part of which constitutes elimination of paternally derived (imprinted) chromosomes, and part the odd non-disjunctional behaviour of the one remaining X.

Think about the genetics of all this. The male *Sciara* passes on to the next generation only those genes (chromosomes) that he inherited from his mother. Think about some of the developmental implications. Sex is determined by events that take place after formation of the zygote: the gonads differentiate according to their chromosomal environment, i.e. the chromosome constitutions of the surrounding cells, rather than their own chromosomal constitution. Think about imprinting. Sometimes it is operational, as in the eventual elimination of all paternal chromosomes from the male germ-line cells; other times it is absent, as in the random elimination of the L chromosomes from somatic cells.

Lastly, in relation to the matters of chromosome inactivation, elimination and imprinting, we come to the scale insects, the coccid bugs (class Insecta, order Homoptera, superfamily Coccoidea). These tiny, apparently insignificant, creatures, many of which do not really look like insects at all, draw attention to themselves only on account of the damage they do to fruit trees and cultivated roses. Some are so small that it is only possible to see their chromosomes by fixing and squashing an entire animal for each microscope preparation. Several different systems of chromosome behaviour are found amongst these animals. Consideration of two of these systems seems warranted here. In both cases the females are perfectly normal and straightforward and the systems are distinguished only by the chromosome behaviour in the male. In the first, all the paternal chromosomes become condensed at blastula stage in the male and remain so in all the somatic cells for the remainder of the animal's life. As a rule, they clump to form a single solid mass of inactive heterochromatin. First meiosis is equational. Second meiosis separates paternal chromosomes from maternal ones. Sperm develop only from cells that have maternal chromosomes. So the situation resembles *Sciara* in the sense that the male only transmits his mother's genes.

The second system goes a step further. Here the paternal chromosomes are eliminated during late cleavage divisions in male embryos. Spermatogenesis involves just a single non-reductional division. Recall our hypothetical step from X inactivation in eutherian mammals to X elimination in bandicoots.

MORE ABOUT INSECTS

Two more relatively simple examples need to be mentioned in the context of this chapter. The first of these is an Australian grasshopper. In certain grasshoppers and crickets, the X chromosome is generally undercondensed in mitoses leading up to male meiosis. Then, during prophase of meiosis, it becomes overcondensed and shows as a compact, heavily staining rod in the meiotic prophase nucleus (Figure 6.4). No other chromosome shows this behaviour. In the female, where there are two X chromosomes, they do not show differential condensation at any stage. Most orthopteroid insects (crickets, grasshoppers and locusts) have XO males, there being no Y chromosome, but in a few species a centric fusion has happened between the acrocentric X and an acrocentric autosome to produce a metacentric **neo-X chromosome**. The unfused member of the autosomal pair then behaves as a neo-Y and in this unfused form it is confined to the male line.

   In such a case, only one arm of the X, that which constituted the original (ancestral) X, is undercondensed in spermatogonial mitoses and overcondensed

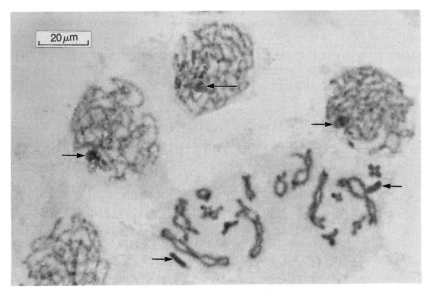

Figure 6.4 Six cells in the first meiotic division from the testis of the locust *Locusta migratoria,* four of them in pachytene and two in diplotene. The precociously condensed X chromosome is indicated by an arrow in each cell. In diplotene the X appears as a short, compact rod-like univalent.

in meiotic prophase. Here we should pause to take note of the difference between this situation and the one that prevails in mammals when we translocate an X onto an autosome and the X imposes its cycle of condensation and inactivation on the autosome to which it is attached. The extraordinary feature of the grasshopper system emerges only when we examine the occasional tetraploid spermatogonia that are not uncommon in orthopteroid testes. These cells have two neo-Xs and two neo-Ys; but only one of the neo-Xs shows heterochromatic behaviour in its X-arm. In principle, this would seem to represent something akin to the behaviour of the mammalian X, and, if so, then we could suppose that at least in the spermatogonia the activity of *one* X is important right through to metaphase of each spermatogonial mitosis; but if two X chromosomes are present, as in tetraploid cells, then a dosage compensation comes into effect and one of the Xs condenses along with all the autosomes. Of course, an interpretation of this kind hangs upon the notion that decondensation and activity go hand in hand.

The second situation that deserves mention relates to the Hymenoptera (bees and wasps). Here the case can be stated quite simply. Bees and wasps have **haploid impaternate males** produced parthenogenetically from unfertilized eggs. Females are all normal diploids. If one counts the chromosomes and measures the DNA in embryonic male nuclei, then one finds the haploid chromosome number and the C amount of DNA. C-values for the group are generally within the range 0.1–0.5 pg. If, however, one examines the somatic cells of adult bees and wasps, then haploid nuclei are not to be found. Even the haemocytes (blood cells) of impaternate males always contain at least the 2C quantity of DNA, and other types of cell show variable degrees of polyploidy ranging from diploid to 32-ploid. We may therefore suppose that a **positive dosage compensation** mechanism operates in genetically haploid males such that at least one additional cycle of DNA replication ensures that a certain minimum (2C) quantity of DNA is established in every somatic cell.

ASCARIS AND CHROMATIN ELIMINATION

So far in this chapter we have considered three events that seem to have formed the bases of a range of strategies employed by animals to deal with certain aspects of bisexuality and male heterogamety (XY or XO sex-determining systems): random inactivation, chromosome imprinting and chromosome elimination. Each of these events could logically be regarded as another expression of the general principle that mechanisms exist for selectively activating or deactivating some parts of a genome without affecting the other parts. Later we shall see this principle in action in relation to DNA replication, polytenization and gene amplification. Here we see it in action in relation to dosage compensation, sex determination and differentiation between germ cells and soma. The chapter ends with a consideration of the most spectacular and bizarre of all chromosomal phenomena, first seen more than 100 years ago by one of the greatest cytologists of all time, Theodor Boveri, the man, incidentally, who, in a completely different context, first made the connection between chromosomal changes and the origin of cancerous tumours.

The object of Boveri's researches was a large (20–30 cm long) parasitic roundworm (Nematoda), *Parascaris equorum* (called *Ascaris megalocephala* in Boveri's time) that inhabits the gut of horses. It is said that he selected this animal because he was interested in early developmental processes. In most ascarid worms the fertilized eggs, many thousands of them, accomplish much of their early development while still in the uterus of the female. They therefore represent an excellent source of material for the study of early embryonic development in these animals.

P. equorum appears to have just two very long chromosomes in each cell following the first cleavage division of the fertilized egg. The middle portions of these chromosomes are more slender and less condensed than the end portions. During metaphase and anaphase of the second division, something quite extraordinary happens. One of these cells is destined to be the 'stem cell' whose descendants will ultimately give rise to all the germ cells of the organism. This cell divides normally, with its two long chromosomes splitting in the usual way and distributing equationally to the two daughter cells. Interestingly, most of the classical descriptions of this division indicate that the chromosomes associate with the mitotic spindle in a manner that suggests attachment to the spindle at many points all along their lengths.

The other cell in the two cell embryo will give rise only to cells that will make up the body of the animal, but not the ovaries or testes. Its division is preceded at prophase and metaphase by the detachment of the more condensed end regions of the chromosomes and break up of the middle regions into 43 (in the male embryo) or 48 (in the female embryo) smaller fragments, each of which then attaches to the spindle by its own independent centromere, splits and then divides equationally into the two daughter cells. The detached ends fail to attach to the spindle and are excluded – eliminated – from the daughter cells. The fate of the detached ends is referred to in the classical literature as **chromatin elimination or diminution**. At the third cleavage division, the cells that have already lost the ends of their chromosomes divide normally with 43 or 48 chromosomes each. One of the other pair of cells divides normally and equationally with its two large chromosomes. The other eliminates its chromosome ends in the same manner as did one of the cells in the second division. The process is then repeated once again, so that at the 16-cell stage there are 14 cells with lots of short chromosomes and two cells with a few long ones. One of the latter divides with elimination. The other sets out on a series of divisions, without any further elimination, that will give rise to all the germ cells of the animal. The sequence of events is shown diagrammatically in Figure 6.5.

The 'Ascaris story' is a particularly satisfying one in the sense that it represents an investigation that has continued and progressed almost without

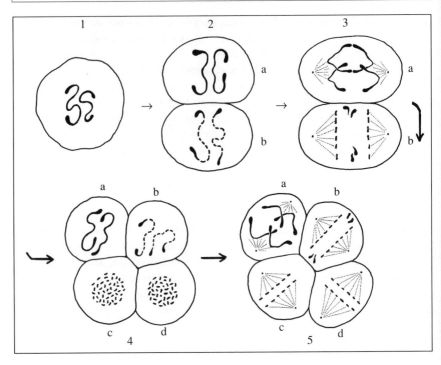

Figure 6.5 The first two and the beginning of the third cleavage divisions in the nematode *Parascaris equorum (univalens)*. The first elimination and fragmentation of the chromosomes takes place as cell 2b prepares for the second cleavage division. Its progeny, cells 4c and d, have numerous small chromosomes and lack the DNA that made up the dumbbell-shaped ends of the two large chromosomes in cell 2b. A second elimination takes place in cell 4b as it proceeds through to the third cleavage division, so that at the eight-cell stage (not shown here) there will be six cells with numerous small chromosomes and two with two large chromosomes, one of which will undergo a third elimination. The cell that remains with two large chromosomes at the 16-cell stage will be the progenitor of all germ cells. The other cells will all contribute to the formation of the body (soma) of the animal.

interruption ever since Boveri's first discoveries. The following points are particularly significant. *Ascaris suum,* the pig nematode, has 48 chromosomes in all its cells at all times, but it too eliminates certain components of these chromosomes during the early cleavage divisions of potential somatic cells. In *Ascaris lumbricoides,* the species that infects humans, the heterochromatic terminal segments of two large chromosomes are eliminated in the same fashion as in *P. equorum,* but the middle regions of the chromosomes do not fragment. *A. suum* eliminates about 22% of its chromosomal DNA. *P. equorum* eliminates about 85%. In both species the germ-line DNA that is eliminated is made up of a very few families of highly repeated short DNA sequences. Two of these have been characterized in *P. equorum*. They consist of 125 and 131 nucleotide pairs respectively and are each repeated about half a million times with the repeats arranged in tandem to serve as a basis for the large heterochromatic blocks that make up the eliminated chromosome ends. In *A. suum* the germ-line chromosomes have sticky ends and show a tendency to aggregate into larger 'plurivalent' chromosomes. The germ-line chromosomes of *P. equorum* are **holocentric**, which means that they have no localized centromere but spindle fibres associate with them all along their lengths. This property is suddenly lost immediately prior to elimination.

The molecular basis of chromatin elimination in *Ascaris* is now quite well understood. It has been investigated, with considerable expenditure of time and resources, not because of a localized interest in the behaviour of chromosomes in nematode worms, but because of a much more basic curiosity about what lies at the natural ends of chromosomes. Normally, chromosome ends are totally unreactive. Chromosomes do not stick to one another, nor do they grow or shorten, except in very special circumstances. On the other hand, if we break a chromosome, the broken ends are highly reactive and will tend to join up with any other broken ends very quickly indeed. Non-reactivity seems to be a property of natural **telomeres**, so the question arises: what is a telomere in molecular terms? *Parascaris* may help to provide answers to this problem, for at one moment it has two stable chromosomes with four telomeres, and at the next it has 48 stable chromosomes with 96 telomeres. What seals the ends of these chromosomes? Are the points of fragmentation defined by DNA sequences that represent incipient telomeres? The story is not yet by any means complete. Suggestions for further reading on the most recent researches on this very exciting topic can be found at the end of the chapter.

Precisely where *Parascaris* fits into the general problem of differential chromosome behaviour is hard to say. Our interpretation of what is accomplished by chromatin elimination depends entirely on our view of the selective pressures and evolutionary pathways that are likely to have been involved. On the one hand, for example, we might say that it proved to be a selective advantage to reduce the amount of chromatin, the overall genome size, in the somatic cells of the organism, perhaps leading to accelerated development through shorter cell cycles. On the other hand, we might argue that it was a selective advantage to retain in the germ line a plurichromosomal arrangement with substantial heterochromatic blocks in order to maintain certain favourable patterns of recombination and segregation. Whatever the advantages, what *Parascaris* does is by no means novel or particularly original. The same general strategy was probably well established amongst the arthropods and the ciliated protozoa long before the days of parasitic nematodes.

Amongst the arthropods, the little copepod crustacean *Cyclops* (2N=22) eliminates more than 50% of its chromosomal DNA from future somatic cells. Cytological studies have shown that the eliminated chromatin derives from the disintegration of distinct heterochromatic chromosomal segments. Some of these are terminal but others are interstitial, which implies a mechanism for removing integrated chromosome segments without impairing the structural integrity of the remaining chromosome. As in nematodes, the elimination process happens early in development somewhere between the fourth and seventh cleavage divisions.

THE PROTOZOAN MACRONUCLEUS

Ciliated protozoa, amongst which the most familiar organisms are probably *Paramoecium* and *Tetrahymena*, are characterized by having two kinds of nucleus, a micronucleus and a macronucleus. Only the micronucleus can undergo meiosis, and it alone is active in promoting sexual reproduction and in transmitting genes from one generation to the next. The macronucleus is a temporary structure that divides amitotically in each round of asexual multiplication but is broken down and reformed whenever the cell undergoes sexual reproduction. The macronucleus has a purely trophic function, providing the transcriptive activity needed for development and growth of the cell. This separation of the nuclear functions within a single cell parallels the soma/germ-line differentiation that characterizes all multicellular eukaryotes.

The macronucleus, as its name suggests, is large and contains much more than the diploid amount of DNA. In *Tetrahymena*, for example, there are about 45 haploid genomes' worth of DNA in each macronucleus and that DNA includes at least 90% of the sequences present in the micronucleus. In another group of ciliated protozoa, known as the hypotrich ciliates, the macronuclear events are more complex. The organism known as *Stylonychia* provides a good example and has been investigated in some detail (Figure 6.6). Here the macronucleus develops through *two* cycles of polyploidization. In the first of these, small cross-banded chromosomes appear in which the bands are clusters of tiny loops

Figure 6.6 Macronuclear development and associated chromosomal and DNA events in the ciliated protozoon *Stylonychia mytilus* (in general according to Ammermann *et al.*, 1974, but with additional interpretations by the author of more recent evidence). After conjugation in this organism the two cells separate and the nucleus that is to form the new macronucleus is just a little larger than the micronucleus (time zero, stage 1). Some hours later the macronucleus swells and its chromosomes become visible (2N = about 280). In stage 3, the chromosomes develop into giant 'polytene' structures with intense DNA synthesis. These chromosomes are not strictly comparable to the polytene chromosomes of insect salivary glands (Chapter 7): their banded appearance is produced by regions of the chromosome where the DNA strands are looped outwards from the chromosome axis (stages 3 and 4 in the diagram). There is then a sudden drop in the DNA content of the nucleus which is caused by the excision and loss of all the DNA in the lateral loops, leaving only fragments of chromosomes that represent parts of the regions between the bunches of loops (stage 5). These remaining regions, together representing only a small fraction of the DNA sequences in the whole genome, are then reamplified in a second period of DNA synthesis, giving rise to the mature and full-sized macronucleus (stage 6). DNA content is in arbitrary units as measured by Ammermann *et al.* (1974) using a microspectrophotometer.

of DNA attached to a thin DNA axis that extends continuously throughout the length of the chromosome. Each loop corresponds in size to about 40 kilobases of DNA. Soon after they are formed, these banded chromosomes break up and lose all their loops. The loop DNA is degraded and lost, resulting in a reduction of more than 95% in the DNA content of the nucleus. The remaining DNA, which is now present in gene-sized pieces each with two new stable telomeric ends, is then re-replicated to generate a macronucleus that has a DNA content equivalent to about 2000 times that of the micronucleus. The main point here is that in the formation of the macronucleus there is a massive elimination of

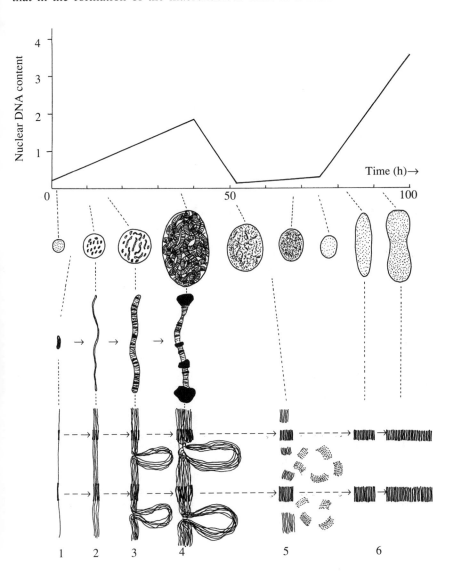

chromosomal DNA, such that the final nucleus with a 4096C DNA content is formed from only 1.6% of the sequences represented in the micronuclear genome.

There is not much room for doubt here with regard to evolutionary pathways and selective forces. It seems most unlikely that the ancestral ciliate genome, as represented by the micronucleus, was small and gradually expanded by more than tenfold to the size of today's micronuclear genome in response to selection for a larger karyotype and an attendant pattern of recombination and chromosome inheritance, and that a system co-evolved for off-loading the extra DNA in the somatic (macronuclear) line. Surely it is much more likely that macronuclear efficiency was the driving force, accomplished by a selective elimination and subsequent reamplification only of those genes that contributed to growth and development.

RECOMMENDED FURTHER READING

Ammermann, D., Steinbruck, G., Berger, L. and Hennig, W. (1974) The development of the macronucleus of the ciliated protozoan *Stylonychia mytilus*. *Chromosoma (Berl.)* **45**, 401–429.

Ballabio, A. and Willard, H.F. (1992) Mammalian X chromosome inactivation and the XIST gene. *Curr. Opin. Genet. Dev.* **2**, 439–447.

Goday, G., Gonzalez-Garcia, J.M., Esteban, M.R., Giovinazzo, G. and Pimpinelli, S. (1992) Kinetochores and chromatin diminution in early embryos of *Parascaris univalens*. *J. Cell Biol.* **118**, 23–32.

Lovell-Badge, R. (1991) X marks the spot. *Curr. Biol.* **1**, 168–170.

Meyer, G.F. and Lipps, H.J. (1984) Electron microscopy of surface spread polytene chromosomes of *Drosophila and Stylonychia. Chromosoma (Berl.)* **89**, 107–110.

Monk, M. and Surani, A. (Eds.) (1990) *Genomic Imprinting. Development,* 1990 Supplement. The Company of Biologists, Cambridge.

Prescott, D.M. 1992. The unusual organization and processing power of genomic DNA in hypotrichous ciliates. *Trends Genet.* **8**, 439–445.

Prescott, D.M., Murti, K.G. and Bostock, C.J. (1973) Genetic apparatus of *Stylonychia* sp. *Nature* **242**, 576 and 597–600.

Riggs, A.D. and Pfeifer, G.P. (1992) X chromosome inactivation and cell memory. *Trends Genet.* **8**, 169–174.

Stoll, S., Schmid, M. and Lipps, H.J. (1991) The organization of macronuclear DNA sequences associated with C(4) A(4) repeats in the polytene chromosomes of *Stylonychia lemnae. Chromosoma (Berl.)* **100**, 300–304.

Sapienza, C. (1990) Parental imprinting of genes. *Sci. Amer.* **263**, 26–32.

Tobler, H. (1986) The differentiation of germ and somatic cell lines in nematodes. In *Germ Line–Soma Differentiation.* W. Hennig (Ed.). Springer-Verlag, Berlin, pp. 1–58.

Tobler, H., Etter, A. and Muller, F. (1992) Chromatin diminution in nematode development. *Trends Genet.* **8**, 427–432.

DNA replication, endomitosis and trophic polyploidy | 7

THE SEMICONSERVATIVE REPLICATION OF CHROMOSOMES

From the molecular standpoint, DNA replication is well understood. A chromosome, whether from a virus, a bacterium or a eukaryote, is fundamentally a very long DNA molecule. Chromosomes replicate semiconservatively. This much we learned from Herbert Taylor's classic experiments on *Vicia faba*, the first of their kind, in which we saw how it is possible to draw conclusions at the molecular level from studies with a microscope. Taylor's first experiment in which he demonstrated semiconservative replication of whole chromosomes was soon complemented by Meselson and Stahl's experiment demonstrating semiconservative replication of DNA. Then followed John Cairns' observations on autoradiographs of bacterial DNA caught in the act of replicating in the presence of [³H]thymidine and the demonstration of circular DNA molecules and replication forks. Taylor extended his observations to Chinese hamsters and discovered that not all chromosomes or even parts of the same chromosome replicate at the same time, so giving us the concept of asynchronous replication. Individual chromosomes were shown to start their replication at different points along their lengths, and this simple observation went a long way towards explaining how a eukaryotic chromosome representing a molecule of DNA 10 centimetres long can replicate as fast as a bacterial chromosome with just 1 millimetre of DNA. The notion of early- and late-replicating segments in the chromosomes of higher organisms was introduced and, as we have already seen, attempts were made to use replication patterns for chromosome identification. Lastly in the general story, chromosome replication was visualized at the molecular level by the application of the autoradiographic technique to DNA isolated from replicating eukaryotic chromosomes. Multiple initiation points and bidirectional replication from each of these points was convincingly demonstrated and a series of investigations was set up in search of relationships between the duration of S-phase, C-value, the numbers and spacings of initiation points and the lengths of individual replicating units. It all happened in less than 10 years. It was an exciting field of cell biology. The hypotheses, based on the newly discovered structure of DNA, were straightforward. The experiments to test them were simple and mostly based on the use of radiolabelled molecular precursors and their detection in newly replicated DNA by autoradiography. The results were all beautifully clear-cut and the people who produced them grew famous overnight.

A diagrammatic representation of Taylor's pioneering experiment on *Vicia* is shown in Figure 7.1. A clear understanding of this experiment is absolutely fundamental to an understanding and full appreciation of the whole range of studies in chromosome replication and behaviour that have occupied the past 35 years. The diagram shows four parallel tracks from left to right. The top one shows the experimental protocol in which thymidine labelled with tritium is put in the water that surrounds *Vicia* root tips for a little more than the duration of the first S-phase. The germinating beans are then transferred to water that has no tritiated thymidine, but to which colchicine has been added to arrest all dividing cells in metaphase, and the colchicine is left there for the duration of the experiment. The second track in the diagram shows three complete cell cycles from telophase 0 to telophase 3, including three S-phases, in the first of which the radiolabelled thymidine is available for incorporation into DNA. The third track follows the appearance of a chromosome and the products of its replication through the three cell cycles. Where the chromosome is surrounded by dots it signifies that it appears labelled with silver grains in the autoradiograph when examined with a microscope. The fourth track represents the chromosome and its products as DNA duplexes that replicate semiconservatively. The two parallel lines at the start are two single strands that make up a double helix. Subsequently, a continuous line represents an unlabelled strand and a dotted line represents a radiolabelled strand. The Y formations represent replicating molecules (chromosomes) and the apex of the Y the replication fork.

THE PRINCIPLES OF TROPHIC POLYPLOIDY

As cytologists with an interest in supramolecular phenomena and the evolution of genomes and chromosomes, it is essential that we reach beyond the molecular formalities of DNA replication, for here is an area of nature where all kinds of strange feats have been accomplished and all kinds of clever strategies have evolved in response to natural selection. Our main concern in this chapter will be with **polyploidy** as a consequence of **endomitosis**, or chromosome multiplication in the absence of cell division. That will be followed in a later chapter with a discussion of gene amplification, which also involves DNA synthesis but of a kind that is quite different from the semiconservative replication of whole chromosomes.

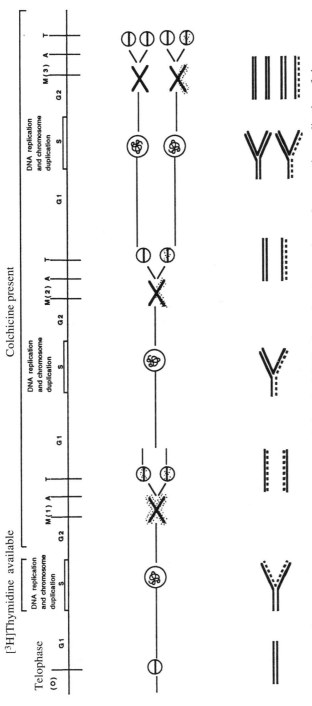

Figure 7.1 'Flow chart' of the classic experiment carried out by J.H. Taylor to demonstrate the semiconservative replication of chromosomes by the combined use of tritiated thymidine, colchicine and autoradiography. The top line of the diagram shows the timing of availability of labelled thymidine and colchicine. Note that the tritiated thymidine is only available to the cells during their first S-phase. The second line of the diagram shows three cell cycles from telophase zero (left) to metaphase/anaphase/telophase 3 (right). The third line of the diagram shows whether or not the chromosomes/chromatids were labelled with silver grains (produced by exposure of the autoradiographic emulsion to beta particles from radioactive thymidine) in the autoradiographs. Dots around the chromatids indicate labelling. The bottom line of the diagram is an interpretative drawing of semiconservative labelling of double-stranded DNA throughout the course of the experiment. The dotted lines indicate half-helices containing tritiated thymidine.

Polyploidy in relation to animals means two things. On the one hand there is the 'genetic' polyploidy that has been discussed in Chapter 5, the evolution and adoption of a polyploid karyotype that is passed on from generation to generation and represents a potent mechanism for the emergence of new species. On the other hand, we have trophic polyploidy, which implies an increase in the number of chromosome sets per nucleus, such that the activity of certain cells in an organism is enhanced in connection with some aspect of that organism's natural history. That is the kind of polyploidy covered in this chapter.

First, a few elementary rules. In large cells, the nucleocytoplasmic ratio is generally shifted towards the cytoplasm, which is to say that big cells have lots of cytoplasmic capacity. However, in organisms with large genomes, the nucleocytoplasmic ratio seems to be shifted the other way: animals with large genomes have less cytoplasm per nucleus than those with small genomes. So one might argue that organisms with small genomes have cells that become polyploid and so increase the number of coding sequences with less risk of loss of cytoplasmic capacity than would be the case in organisms with large genomes. In accordance with this argument, most examples of highly polyploid cells are to be found in animals with small genomes.

One might also argue that, as the cell size increases, the surface area to volume ratio will shift towards volume and, so long as the cell remains more or less round in shape, diffusion paths will become longer and the proportional surface area available for exchange will diminish. Large cells with polyploid nuclei are therefore likely to be slow mass production units, lacking in reactivity and versatility. The exception that helps to prove the rule is that large neurones, which are frequently highly polyploid, characteristically have cell membranes that are richly folded, so greatly increasing their surface area.

Trophic polyploidy is normally a consequence of chromosome duplication without accompanying nuclear division. Since this general phenomenon presents us with such an interesting array of extraordinary situations, I have chosen to examine some specific cases, each of which raises its own special problems and questions, whilst helping to establish an overall picture of the adaptive significance of polyploidy and endoduplication of chromosomes.

SOME EXAMPLES OF POLYPLOID CELLS

The simplest example of polyploidy is found in mammalian liver. Here we find a mixture of cells that are mainly diploid, tetraploid and octaploid. The levels of ploidy can be determined by eye in histological preparations on account of the different sizes of cell nuclei. Chromatin distribution is the same in all nuclei.

There is much more to be learned about polyploidy from insects, and especially from their ovaries. The ovaries of all insects consist of bundles of tubes arranged exactly like a bunch of bananas, with the tubes congregating into a single oviduct (corresponding to the stalk of the banana bunch) at the posterior end. Each tube is called an **ovariole**.

In **panoistic** (Greek: *pan*, all; *oon*, egg) ovaries (Figure 7.2a), each ovariole consists of an uninterrupted line of developing eggs that increase in size from front to rear. The wall of the ovariole consists of two or three layers of **follicle cells**, which are for the most part small and diploid. Each egg in this situation develops and grows on its own, obtaining the necessary materials from or through the follicle cell. The cockroach provides a good example of such an ovary.

In **meroistic** (Greek: *meros*, part; *oon*, egg) ovaries (Figure 7.2b) the egg is assisted in its growth and development by nutritive or 'nurse' cells, as well as by the surrounding follicle cells. Meroistic ovaries can be **polytrophic**, which means that each developing egg has its own independent group of nurse cells. Egg and nurse cells make up a single trophic system and there are many trophic systems in each ovariole, hence *poly* trophic; or they can be **telotrophic** (Figure 7.2c), in which case there is a continuous string of eggs as in the panoistic ovary, but each of these is supplied with nutritive materials via a tube that connects it to a mass of nurse cells that forms a common trophic region at the anterior end (hence *telo* trophic) of the ovariole.

All nurse cells are more or less polyploid. In polytrophic ovaries, nurse cells and egg have a common origin. They all derive from one original cell. In earwigs (*Forficula*), for example, there is just one nurse cell to each oocyte. Both are derived from the division of a single cell, following which one of the daughter cells became polyploid and the other became an egg. In butterflies of the genus *Vanessa*, there are three nurse cells to each oocyte, so two divisions gave four cells, of which one became the egg and three went polyploid. In water beetles (*Dytiscus*), there are 15 nurse cells to each oocyte, 16 cells in all resulting from four divisions.

In *Notonecta*, the back swimmer, a relative of the water boatman and a member of the insect order Hemiptera, the ovary is telotrophic, and it has been estimated that a trophic region comprising some 20 000 nurse cells serves an ovariole containing up to 20 developing eggs. The mechanism whereby the trophic region is connected to individual eggs is absolutely fascinating and represents one of the most spectacular known examples of an intracellular transport system. It has been described by Stebbings and Hyams in their 1979 book, *Cell Motility* (Longman), and the reader will certainly find it rewarding to explore this particular topic in more depth.

Three points are important in relation to polyploidy in the nurse cells of insect ovaries. First, all nurse cells in a group are in cytoplasmic continuity with one another and with the oocyte across cytoplasmic bridges, the patterns of which provide clues to the sequence of division that originally led to the formation of the two-, four-, eight- or 16-cell group. Yet in spite of this cytoplasmic continuity, each nurse cell replicates its DNA independently and out of synchrony with all the others. Secondly, each nurse cell eventually reaches a ploidy level of

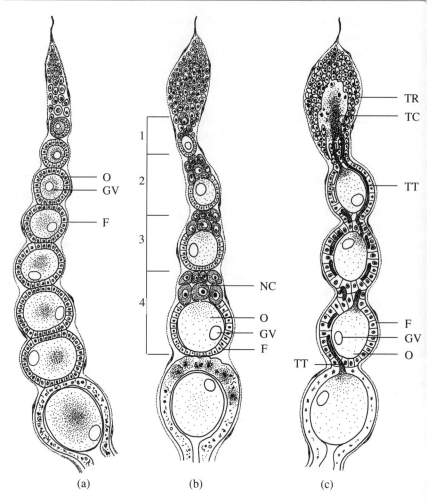

Figure 7.2 Three types of insect ovariole. (a) Panoistic (Greek: *pan*, all; *oon*, egg) ovariole, such as is found in cockroaches, consisting only of a line of growing oocytes (O), each enclosed in a layer of follicle cells (F), with the smallest oocytes at the anterior end. None of the cells in an ovary of this kind are conspicuously polyploid. GV, germinal vesicle or oocyte nucleus. (b) A meroistic (Greek: *meros*, part; *oon*, egg) polytrophic ovariole, such as is found in flies and beetles, consisting of a series of compartments (numbers 1–4), each of which contains one developing egg or oocyte joined through a system of cytoplasmic bridges to seven (or one, three or 15) highly polyploid nurse cells (NC). (c) A meroistic telotrophic (Greek: *telos*, end; *trophe*, nourishment) ovariole, as found in most hemipteran bugs, consisting of a continuous line of growing oocytes each connected by a trophic tube to an anterior trophic region that consists of numerous polyploid nutritive cells discharging nutritive materials into the trophic core from which the tubes lead backwards to the oocytes: TR, trophic region; TC, trophic core; TT, trophic tubes.

at least 1000×, although this has proved hard to determine accurately because of the very large size and density of these nuclei. The largest nurse cell nuclei in polytrophic ovaries are up to 100 µm in diameter. Most of their chromatin is finely granular, but the biggest nuclei often contain star-shaped aggregates of chromatin that are so compact as to have an almost crystalline appearance. Thirdly, all these nurse cells synthesize RNA intensively.

So we have nuclei that go through waves of DNA synthesis to reach ploidy levels in excess of 1000-fold and at the same time synthesize RNA using much, but certainly not all, of their nuclear DNA as a template. What is the purpose of all this? To provide the oocyte with certain essential materials? What materials? What kinds of RNA? In the sense that there are three kinds of cell involved, the oocyte, the follicle cells and the nurse cells, all of which are geared to the production of eggs, to what extent is each specialized in terms of its own transcriptive programme and translational potential?

The situation in telotrophic ovaries is basically the same, but it does show some interesting peculiarities that have yet to be explained. In *Notonecta*, for example, ploidy levels for the nurse cells range from two to about 64 and there are about 20 000 nurse cells and 20 oocytes per ovariole. The ratio of polyploid nurse cells to oocytes is therefore in the region of 1000:1. If the average level of ploidy in the nurse cells is 16×, then each oocyte may have up to 16 000 copies of every gene or gene complex available to it for the synthesis and supply of nutritive materials. Each nurse cell nucleus has a large and compact mass of inactive chromatin that must account for at least half the nuclear DNA, and each has a single very large nucleolus. The trophic tubes that lead from the trophic region to the oocytes are packed with ribosomes on their way to the oocytes. Could it be that the nurse cells in telotrophic ovaries are specialized for the synthesis of ribosomes and they do very little else?

Then we have some truly remarkable examples of polyploidy. The highest level of polyploidy yet recorded is found in a nucleus of a giant neurone in the abdominal ganglion of the mollusc *Aplysia*. One of the cells in this ganglion reaches a size of up to 1 mm in diameter and has an ellipsoidal nucleus that may be half a millimetre long and 100 µm wide. Microphotometric measurements of this nucleus have shown that its DNA increases by a series of doubling steps until it contains 75 000 times the haploid amount for the species. The reasons behind this massive level of ploidy are not understood.

The mosquito, *Culex pipiens,* has a diploid chromosome number of 6. Yet in the late larval or early pupal intestine, where many of the cells are degenerating, mitoses have been described where there are 6, 9, 12, 18, 24, 36 or 72 chromosomes. Now obviously, some of these numbers reflect divisions of cells that are already normal polyploids, that is, the numbers in the 6× series; but what of the 9s, the 18s, the 36s and the 72s? They must have involved, at least in the first stage, a replication of one set of chromosomes but not of the other.

There is another odd situation in the ganglionic cells of *Drosophila* larvae. Whole sets of metaphase chromosomes from first instar larvae have half the

DNA content of those from third instar larvae, but in both cases the number of chromosomes on the spindle is the same. One can reasonably assume that the situation in third instar cells is one in which we have a 'multistranded' condition where each chromosome at metaphase is correspondingly thicker than its double stranded (two-chromatid) counterpart in first instar cells. How, one might ask, do the multistranded chromosomes achieve equational distribution of their sister strands at anaphase?

A final example that has to be taken carefully into account whenever we consider trophic polyploidy amongst insects concerns mealy bugs. They are most commonly encountered amongst the white powdery deposit that sometimes covers old potatoes, and they are so small that one has no option but to squash the entire animal in order to make preparations of its chromosomes! For many years it has been known that the paternally inherited set of chromosomes in mealy bugs (*Planococcus*) and other members of the coccid family remains genetically inactive in the male and is recognizable in male somatic cells as a compact mass of darkly staining chromatin. Certain cells in the mealy bug undergo endomitosis with visible and countable chromosomes. Where this is the case, it is clear that it is only the maternally inherited set of chromosomes that is duplicating, while the paternally inherited chromosomes remain haploid and cytologically distinct. Both sets of chromosomes replicate to produce a polyploid situation in the somatic cells of the female, although the paternally derived set does replicate later than the maternal one. How can the cell discriminate? There is more about this in Chapter 6 in the context of chromosome inactivation.

POLYTENE CHROMOSOMES

One particular kind of trophic polyploidy requires special mention and more detailed treatment: the giant polytene cells of the salivary glands and other tissues in the larvae of two-winged flies (Diptera: midges, mosquitos, house flies, fruit flies and black flies). **Polytene chromosomes** (Figure 7.3) are often referred to as salivary gland chromosomes. The term salivary can be misleading. The glands do not produce saliva in the normal sense, but are functionally more like the silk glands of spiders in that they produce large amounts of a sticky secretion, a mucopolysaccharide, that is mainly used for constructing feeding nets, pupal capsules and, in the case of blackfly larvae that live in fast-flowing water, 'life lines' to help the larva cling to the undersides of stones and prevent it being washed away downstream. In polytene cells we undoubtedly have situations where a high level of effective polyploidy is associated with the production of relatively massive amounts of certain secretory proteins.

The study of polytene chromosomes began over 100 years ago in 1881, when Professor Balbiani of Paris University recognized and described these structures in the nuclei of salivary gland cells from the larvae of the midge *Chironomus plumosus*. Balbiani's description consists of just four short pages of French with

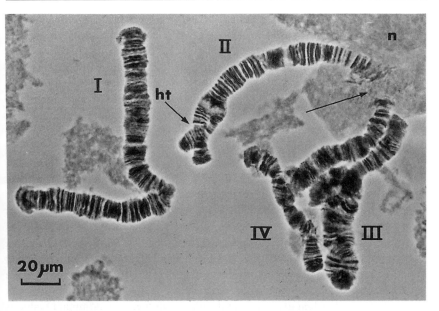

Figure 7.3 Polytene chromosome complement of a salivary gland nucleus from a fourth instar larva of *Chironomus pallidivittatus*. The second of the four chromosomes (II) carries a nucleolus (n) and has a heterozygous, non-pairing, region at one end (ht). The arrow indicates the gap produced in the chromosome by the nucleolus. The shortest chromosome (IV) has two large puffs, the Balbiani rings. Note that the haploid chromosome number for this animal is 4, and since there are only four polytene chromosomes in this originally diploid cell, it follows that each visible chromosome represents the polytenized form of a fused pair of homologous chromosomes. (Picture kindly provided by Dr U. Grossbach and reproduced here with the permission of John Wiley & Sons, Chichester and New York.)

a few diagrams. It is florid and subjective, but in most senses absolutely correct. Research on polytene chromosomes went on steadily over the next 50 years with a growing awareness of their significance but really very little understanding of their structure. Then, in 1930, one investigator (Kostoff) suggested that the cross-banded appearance of the chromosomes might have some relationship to genes. Another investigator (Painter) looked at the polytene chromosomes of *Drosophila melanogaster* whose genetics was already well understood. He made chromosome maps and recognized that translocations and inversions were clearly identifiable as corresponding rearrangements of the chromosomes' cross-bands. Then two more of that outstanding school of pioneering cytogeneticists of the 1930s, Heitz and Bauer, produced what was the first clear statement that the objects inside the nuclei of dipteran salivary gland cells were chromosomes and that, as a rule, they are present in the characteristic haploid number for the species. The apparent haploidy is attributed to the intimate pairing of homologous chromosomes.

Around 1935, Calvin Bridges took over the study of *Drosophila* polytene chromosomes and mapped them in detail. An example of a polytene map is shown in Figure 7.4, that being the map of chromosome 4 in *Chironomus tentans* published by Wolfgang Beermann in 1952. Bridges estimated the total number of cross-bands in *Drosophila* polytenes and put forward the hypothesis that the chromosomes are greatly extended in length, they have duplicated many times over and the products of these duplications have remained side by side, stuck to one another and in perfect register with one another throughout the length of the chromosome. This is why the chromosomes were called polytene (many-stranded).

From 1935 through to about 1950 there was the kind of healthy confusion that is the hallmark of science that is really happening with discovery just around the corner. We have numerous detailed descriptions of structure, of rearrangements associated with genetic experiments, of banding patterns, distributions of 'landmarks', swellings, nucleoli and other features that make up the complex linear differentiation of the chromosomes and all kinds of suggestions as to their functional significance. Let us make no mistake, these chromosomes were a source of immense excitement and targets for a tremendous amount of work throughout an era when genetics and development were the dominant fashions of biological science, an era that led up to and ended with the recognition, in 1952, of DNA as *the* genetic material and a major structural component of a chromosome.

The astonishing thing is that polytene chromosomes, more than 100 years

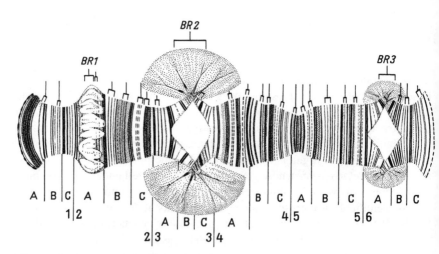

Figure 7.4 Map of the polytene chromosome number 4 of *Chironomus tentans* (modified by Pelling, C., *Chromosoma (Berl.)* **15**, 71–122, 1964). BR1–3, the three Balbiani rings. The chromosome is represented in this map as having more than 160 bands.

after their discovery, are still right at the centre of cytogenetic research: they still dominate the chromosome journals and they still offer a fitting medium for one advance after another in our understanding of chromosome biology.

Drosophila melanogaster has a haploid chromosome number of 4. Chromosome 1, often referred to as the X chromosome, is the largest, and chromosome 4 the smallest. In large salivary gland cells from mature larvae, the nuclei are 2048-ploid. To achieve this level of polyploidy, starting with a cell that is diploid (2-ploid), each chromosome must undergo ten rounds of replication to make 1024 copies of itself. Homologous chromosomes remain fused with one another, as do all the products of their replications, so that at the end of the process, each of the four visible chromosomes has 2048 strands or **chromonemata**. This figure applies only to the largest of the salivary gland nuclei in *D. melanogaster*. In other tissues from the same species, maximum levels of polyploidy are different. In *D. virilis*, for example, it has been estimated that the nuclei of the gastric caecum are 256-ploid, whereas those of the Malpighian tubules are 128-ploid.

The establishment of polyteny proceeds in a series of distinct steps. Each of these represents an S-phase. By examining patterns of incorporation of tritium-labelled thymidine into polytene chromosomes as they replicate it has been possible to show that different regions of the chromosomes replicate at different times. This replicative activity is temporally ordered within as well as between chromosomes, and a definite **replicative chronology** exists for the entire chromosome set. The point at which a particular segment of a chromosome replicates in the replication cycle remains the same even after that segment has been translocated out of its normal position into another chromosome. Specifically, we may suppose that individual bands or groups of bands on polytene chromosomes may behave as independent units of replication within a particular chromosome set. We must also understand, however, that replicative chronology is a characteristic of the whole karyotype. Positional changes within that karyotype will have no effect on the sequence of individual units with respect to one another, but losses or additions of new or different replication units may naturally be expected to produce changes in the overall replication sequence.

SELECTIVE DNA REPLICATION

The main point that has been developed so far in this chapter is that all the DNA in the nucleus is not replicating at once. Some parts are switched on and some are not at any one time. We have already seen examples of this general phenomenon: nurse cells replicate their chromosomes independently of one another even though they share a common cytoplasm, maternal chromosome sets behave differently from paternal ones, there is asynchronous replication during S-phases between mitoses. Now let us look at a few specific situations in polytene cells where DNA synthesis in one family of genes or DNA sequences seems to be entirely out of phase with all the rest of the genome.

If we look amongst the midges that belong to the family Sciaridae, genus *Sciara*, we find that in the polytene chromosomes of these animals there are certain cross-bands that swell up and become very conspicuous immediately before metamorphosis, and these cross-bands are the only remaining sites of DNA synthesis. Throughout most of the larval life of the animal these bands participate in the series of orthodox replications that lead to polytenization of the chromosome, but late in larval life these same regions become structurally differentiated and make extra DNA at a time when normal polytenization has ceased. The extra DNA is made in a series of steps such that, when one measures the amount of DNA in a 'DNA puff' region before the extra DNA starts to accumulate and expresses this as 'p', then the increase in DNA as the puff grows goes stepwise from p–2p–4p–8p–16p. The level at which this extra DNA synthesis stops is characteristic of each particular puff in a given species.

The remarkable thing about DNA puffs is that their DNA remains integrated into the chromosome, which must lead us to ask some more searching questions. Is the extra synthesis merely an indication of a particularly late-replicating region of the chromosome, essentially a region that reaches the final level of polyploidy much later than the rest of the chromosome? Or is it an indication of a selective 'superpolytenization' of one particular band or gene complex? Are the DNA puffs of any developmental significance, or are they merely expressions of degeneracy in a terminal cell that has almost outlived its usefulness? All the evidence currently available points to the conclusion that DNA puffs represent hyperactive loci that are specifically concerned with the production of materials that are important for some aspect of late larval development and/or pupation.

In the present context they provide a good example of one small component of a genome that replicates selectively through several rounds when all else, including cross-bands in the immediate neighbourhood, has long since finished.

The other two features that I wish to mention here concern *Drosophila melanogaster*. In the mitotic prophase chromosomes of *D. melanogaster*, certain regions of the chromosomes are compact and densely staining and their sister chromatids tend to stick tightly together right through to anaphase. These regions are said to consist of **heterochromatin** and to be **heterochromatic** at mitotic prophase and at certain other times. The word heterochromatin simply means chromatin that is different. The distribution of heterochromatin in *D. melanogaster* is shown in Figure 7.5. It comprises the proximal one-third to one-half of chromosome 1 (X), the middle 20% of chromosomes 2 and 3, about half of the little fourth chromosome and all of the Y chromosome. In polytene nuclei from salivary glands, these heterochromatic parts are all fused together into a relatively structureless mass of chromatin, the **chromocentre**, which is much smaller than might be expected. From the arrangement of heterochromatin on chromosomes 1, 2, 3 and Y one can visualize that the cytological

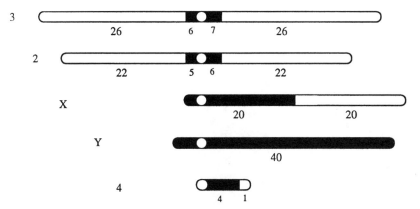

Figure 7.5 Relative proportions in terms of millions of base pairs and distributions of heterochromatin (solid black) and euchromatin (unshaded) in the mitotic chromosomes of *Drosophila melanogaster*. Note how there is heterochromatin surrounding the centromeres on all chromosomes and all of the Y, most of the fourth and half of the X are heterochromatic, leaving four entire chromosome arms and one long half-arm (the X) as euchromatin.

arrangement in the polytene nucleus will be one in which five long chromosome arms and two very short arms will project from the chromocentre (Figure 7.6), reflecting the fact that chromosomes 2 and 3 are metacentric, so contributing two arms each, chromosome 1 is telocentric, contributing one arm, and chromosome 4 is so small that its arms will not be distinguishable.

Deep in the middle of the chromocentre lies a small compact mass that is known as the alpha-heterochromatin and this is surrounded by a larger mass of more loosely textured material that is called the beta-heterochromatin. The alpha-heterochromatin represents most of the mitotic heterochromatin. The beta-heterochromatin represents some relatively small parts of the mitotic chromosomes that lie at the junctions between the main blocks of heterochromatin and the other parts of the chromosomes. Why then is the alpha-heterochromatin so very small in the polytene cell if it corresponds to all of the Y, a third of chromosome 1 and a fifth of chromosomes 2 and 3? The reason is simple: it does not replicate at all.

The plot thickens when we look at the heterochromatic end of the X chromosome. In this region we find the gene complex that codes for ribosomal RNA, in cytological terms the **nucleolus organizer**. Another nucleolus organizer is situated on the short arm of the totally heterochromatic Y chromosome. So the ribosomal gene sequences on both X and Y chromosomes are firmly embedded in material that fails to replicate in polytene cells. Nonetheless, polytene cells have conspicuous nucleoli, which suggests strongly that there must be more than just the diploid equivalent of nucleolus organizers present.

It was against this background that someone decided to measure the relative amounts of the DNA that make up the genes for ribosomal RNA (rDNA) and

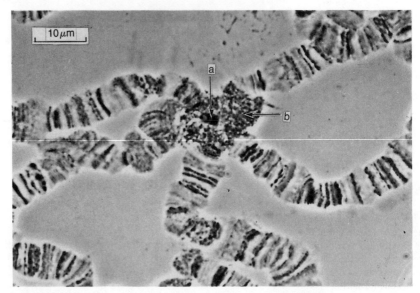

Figure 7.6 *Drosophila virilis* polytene chromosomes showing the alpha-heterochromatin (a), the beta-heterochromatin (b) and the bases of the five chromosome arms that consist mostly of euchromatin (see Figure 7.5). (Reproduced here with the kind permission of Gall, J.G., Cohen, E.H. and Polan, M.L., *Chromosoma (Berl.)* **33**, 319–344, 1971.)

chromosomal DNA in diploid and polytene cells from *Drosophila*. Were the ribosomal genes polytenized and, if not, then might this have something to do with the fact that they were sequestered within a most unusual DNA landscape consisting of heterochromatin that did not participate at all in the polytenization event? The approach to this question was enhanced by the possibility of looking at flies that had two X chromosomes (normal XX females) and those that had just one (XO males) and asking whether the level of polytenization of X-linked genes was proportional to the number of copies of these genes that were present at the start.

The results were astonishing. First, they showed that, in polytene cells, the rDNA is under-replicated with respect to the remainder of the non-heterochromatic chromosomal DNA. The latter does ten rounds of replication; the rDNA does seven or eight. Expressed in another way that reflects the actual observations that were made, 0.4% of the nuclear DNA is rDNA in normal diploid tissues. The corresponding value for polytene cells was 0.1%. Ten rounds of replication would make 1024 copies (starting from one template); eight rounds would make one quarter of that amount, hence, it could be argued, the reduction from 0.4 to 0.1%.

Moreover, it was shown that, whereas in diploid cells the amount of rDNA was strictly in proportion to the number of X chromosomes that carried nucleolus organizers, this was not so for polytene cells. In the latter case, the amount of rDNA is the same whether there be one (XO) or two (XX) X chromosome nucleolus organizers in the cell.

The situation can be explained if one supposes that the nucleolus organizer of XO polytene cells undergoes an additional round of replication so that in the late larva they have as many ribosomal genes, as much rDNA, as the XX cells. On the other hand, one might just as well suppose that only one nucleolus organizer is ever polytenized in those species where the nucleolus organizer is embedded in heterochromatin and is therefore allowed a measure of independence with regard to its behaviour during the process of polytenization. The question might be summed up as compensation versus competition. How might we distinguish between the two possibilities? We shall return to this matter again in a related context in the next chapter.

There then are some of the complexities of polytene cells. What limits the overall level of polyploidy reached by these cells? What molecular features of chromosome bands determine their place in the replicative sequence? Why does heterochromatin fail to replicate? Why do the ribosomal genes behave so differently from the remainder of the chromosome, and how is their 'dosage compensation' accomplished in flies that have different complements of X chromosomes?

Of course there are other aspects of polytene chromosomes that have kept them at the centre of the stage with regard to research on programmed gene expression in differentiation and development, but that is not the subject of this chapter. Here we have been concerned mainly with DNA replication and the establishment of trophic polyploidy. We have seen the replication of the entire genome, of one set of chromosomes but not another, of one type of chromatin but not another, of different parts of chromosomes in relation to each other and of specific gene families independently from all others. In the next chapter we shall consider the independent behaviour of certain gene families as adaptive responses to specific selective forces, in the phenomenon of **gene amplification**.

RECOMMENDED FURTHER READING

DNA puffs

Glover, D.M., Zaha, A., Stocker, A.J., Santelli, R.V., Pueyo, M.T., de Toledo, S.M. and Lara, F.J.S. (1982) Gene amplification in Rynchosciara salivary gland chromosomes. *Proc. Natl. Acad. Sci. USA* **79**, 2947–2951.

More about *Drosophila* heterochromatin

Endow, S. and Gall, J.G. (1975) Differential replication of satellite DNA in polyploid tissues of *Drosophila virilis. Chromosoma (Berl.)* **50**, 175–192.
John, B. and Miklos, G. (1988) *The Eukaryotic Genome in Development and Evolution.*

Allen & Unwin, Boston, Sydney and Wellington.

Miklos, G. and Cotsell, J.N. (1990) Chromosome structure at the interfaces between major chromatin types: *alpha*- and *beta*-heterochromatin. *BioEssays* **12**, 1–6.

Historic

Taylor, J.H., Woods, P.S. and Hughes, W.L. (1957) The organization and duplication of chromosomes as revealed by autoradiographic studies using tritium-labeled thymidine. *Proc. Natl. Acad. Sci. USA* **43**, 122–127.

The nucleolus organizer and gene amplification

In this chapter we shall focus on just one gene complex, the nucleolus organizer, and examine it closely. The reason behind this strategy is simple. The nucleolus organizer was the first major gene complex to be investigated in detail at the molecular level and we know more about it than almost any other gene complex. Moreover, we can use the nucleolus organizer to introduce certain basic principles that are of widespread genetic and evolutionary significance.

BASIC MOLECULAR BIOLOGY

The nucleolus organizer is a specific site on a specific chromosome that has a nucleolus associated with it at certain stages in the cell cycle. In cytological terms, it may be identified by its attached nucleolus or it may appear as a non-staining gap on the nucleolus organizing chromosome. In functional terms, a nucleolus organizer is a chromosomal locus that is primarily devoted to the synthesis of the precursors of ribosomal RNA. Accordingly, one strand of a substantial part of the DNA that resides at this locus in a given organism is complementary in its nucleotide sequence to the 18S and 28S **ribosomal RNA (rRNA)** molecules of the species in question. The nucleolus itself is an agglomerate of nascent rRNA precursor molecules and part-assembled ribosomal subunits.

In all eukaryotes the nucleolus organizer consists of multiple repeats, arranged in tandem, of the sequences for 18S and 28S rRNA interspersed with spacer sequences. The spacers may be transcribed, but their transcripts are never included in the mature ribosome. As a rule, the DNA sequences for 18S and 28S rRNA are closely tandemly linked, with only a short region of **spacer DNA** separating them, and the individual 18S+28S blocks are separated from one another by longer stretches of spacer, most of which is not transcribed. The 'ribosomal gene', usually referred to as **ribosomal DNA (rDNA)**, is a repeated

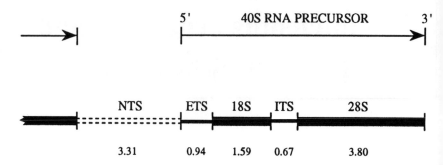

Figure 8.1 Schematic representation of the basic ribosomal gene repeat unit in the frog *Xenopus laevis*. The limits of the repeat are from the beginning of the non-transcribed spacer region (NTS) on the left to the end of the 28S gene on the right. The molecular sizes of each region are given in kilobases along the bottom line of the diagram. *At 3 kb to 1 μm of DNA, how do these dimensions compare with the observed dimensions of transcription units and spacers as seen in Figures 8.6 and 8.9?* The limits of the region transcribed into a 40S precursor ribosomal RNA molecule are shown in the top line of the diagram. ETS, external transcribed spacer; ITS, internal transcribed spacer.

gene with tandem repeats arranged and clustered into one or a few distinct regions of the chromosome set (Figure 8.1). At a single locus we may find fewer than 20 repeats, as, for ·example, in the protozoon *Tetrahymena,* or several thousand repeats, as in the large genomes of some of the tailed amphibians (newts and salamanders). One of the best-known nucleolus organizers is that of *Xenopus laevis*. Each individual nucleolus organizer in this species has between 100 and 500 copies of the rDNA unit.

Why do we know so much about the nucleolus organizers of *Xenopus laevis*? The answer is clear, if a little complicated, and it will be appropriate and helpful to give it immediately. Like other amphibians, and many other kinds of organism besides, *Xenopus* **amplifies** its ribosomal genes in the earliest stages of oogenesis, so that each oocyte comes to contain several thousand extra-chromosomal nucleolus organizers, each of which takes the form of a ring of rDNA that includes many repeats of the 18S + 28S + spacer unit. The amplification manifests itself as a massive synthesis of DNA mainly during pachytene, when up to 30 pg of rDNA accumulates in the oocyte nucleus. The consequences of the amplification are evident in the thousands of extra-chromosomal nucleoli that are present in the oocyte nucleus during its meiotic prophase (see Figure 9.3).

Now a pachytene nucleus from *X. laevis* may be expected to contain four times the C-value equivalent of chromosomal DNA. C for *X. laevis* is 3.2 pg, so the oocyte nucleus will have 12.8 pg of chromosomal DNA. The presence in the same nucleus of 30 pg of amplified rDNA means that more than two-thirds of the entire nuclear DNA in the *Xenopus* oocyte nucleus is pure rDNA. Twenty or thirty years ago, before the days of recombinant DNA technology, *Xenopus* oocytes were regarded as a molecular goldmine from which it was possible to isolate large quantities of this one particular gene complex for subsequent detailed molecular analysis. In the 1960s and 1970s it was one of the most rewarding systems available to the molecular biologist. Hence our rather comprehensive knowledge on the subject.

The underlying usefulness of gene amplification and the multinucleolate condition is well understood. Essentially it enables the oocyte, which is only tetraploid with respect to its *chromosomal* nucleolus organizers, to synthesize as much ribosomal RNA in a few months as would be made by a normal diploid cell in about 450 years! This rRNA, estimated to amount to over 4 *milli*grams per mature egg, is incorporated into the ribosomes that serve in protein synthesis from fertilization through to the hatched and feeding tadpole stage.

IDENTIFYING NUCLEOLUS ORGANIZERS

How does one recognize a nucleolus organizer? The only immediately obvious clue is the presence of a **secondary constriction** on a mitotic metaphase chromosome (the centromere is usually marked by the primary constriction) (Figure 8.2). However, not all secondary constrictions signify nucleolus organizers. There is one on the long arm of chromosome 1 in *Homo sapiens* and there are certainly no ribosomal genes in this position. There are five sites of clusters of ribosomal genes in *Homo* and all of these are in the short arms of the D- and G-group chromosomes (numbers 13, 14, 15, 21 and 22) (Figure 8.2 insert).

What are secondary constrictions? Is there something molecularly peculiar about the chromatin in the region of the nucleolus organizer? Probably not, since the constrictions generally disappear in the course of mitosis and they are quite uncommon in *meiotic* metaphase. Do they represent regions of chromatin where transcription stops later in the cell cycle than elsewhere on the chromosomes and where associated RNA polymerase molecules interfere with the changes in chromatin structure that must surely accompany prophase and metaphase condensation? This is not unlikely, although it still leaves us with the question of why nucleolar transcription should continue when all else has ceased. Or are constrictions just that in the real sense of the word: regions where there has been a local delay in chromosome condensation, a strangulation, caused by tardy dispersal of quite tightly compacted nucleolar material?

Another way of looking at nucleolus organizers is by *in situ* **nucleic acid hybridization**. This approach most commonly begins with quite ordinary cytological preparations in which cell nuclei have been 'fixed' in the histological sense and then stuck to a microscope slide by squashing or smearing of tissue or by some more or less sophisticated but nonetheless standard cytological technique. The preparation is treated in such a way as to denature (make single-stranded) the DNA of the nuclei and chromosomes without removing it from the

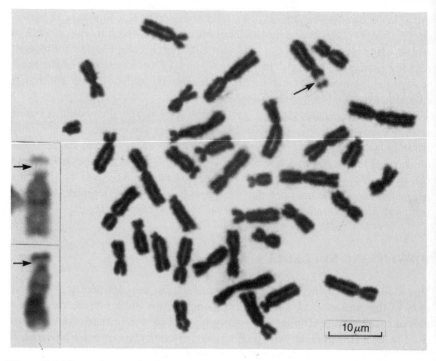

Figure 8.2 Photomicrograph of an entire mitotic metaphase chromosome set from *Xenopus laevis* (2N = 36), showing a single chromosome with a prominent gap (secondary constriction) in the middle of its short arm (arrow) marking the site of a nucleolus organizer. Insert figures show examples of D-group chromosomes 13 (upper) and 15 (lower) from a normal human. The arrows indicate the non-staining gaps that mark the sites of nucleolus organizers on the very short arms of these chromosomes.

slide. A brief dipping in weak sodium hydroxide or gentle heating to about 70°C is the most common method. The preparation is then incubated in a solution of a radioisotopically labelled, single-stranded nucleic acid (the **probe**) that is complementary in nucleotide sequence to a particular chromosomal gene complex (the **target**). The labelled probe anneals specifically to its complementary target sequence(s) and the position of the **hybrid** molecule is determined by **autoradiography**. Nucleolus organizers are easy to locate in this way, mainly because they are relatively large gene complexes, big targets, and they therefore bind relatively large amounts of the labelled probe. Moreover, the probe has always been easy to come by: even though there might be technical or economic difficulties in making or obtaining labelled rDNA, ribosomal RNA is still the most abundant species of RNA in eukaryotic cells and it is naturally single-stranded and easy to label with radioisotopes. An example of a nucleolus organizer located by *in situ* hybridization with a tritium-labelled rDNA probe is shown in Figure 8.3.

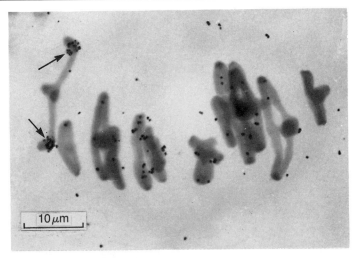

Figure 8.3 Photomicrograph of a first meiotic metaphase/anaphase from the testis of a salamander called *Plethodon cinereus*. The preparation has been treated briefly with dilute sodium hydroxide to denature the chromosomal DNA, then incubated in a solution of tritium-labelled ribosomal RNA and covered with a 'nuclear track emulsion'. The result, after suitable exposure and development of the emulsion, is an autoradiograph in which there are distinct clusters of small black silver grains over each location on the chromosomes where the radiolabelled ribosomal RNA has bound to DNA with a complementary sequence, i.e. locations where there are nucleolus organizers (NORs). Two NORs are indicated by arrows occupying positions near the centromeres on the short arms of a pair of medium-sized submetacentric chromosomes. (Reproduced with the permission of Springer-Verlag from Macgregor, H.C., Vlad, M. and Barnett V., *Chromosoma (Berl.)* **59**, 283–299, 1977.)

A quick method for locating nucleolus organizers involves incubating cytological preparations on microscope slides in a 50% solution of silver nitrate. Under the right conditions of incubation time and temperature, a black grain of metallic silver forms immediately over the nucleolus organizer and is easily located by ordinary light microscopy (Figure 8.4). The one disadvantage of the technique is that it does not stain all nucleolus organizers. Silver staining of metaphase chromosomes is related to transcriptive activity of the nucleolus organizer at the previous interphase. This conclusions stems from a most interesting experiment that involved artificially made mouse–human hybrid cells.

It is known that mouse–human hybrid cells, generated by the artificially induced fusion of cells in tissue culture, synthesize only mouse ribosomal RNA. In such cells, only the nucleolus organizers on the mouse chromosomes stain with silver. The human nucleolus organizers do not stain in hybrid cells, although they do stain in the normal diploid cells of the hybrid's human parent.

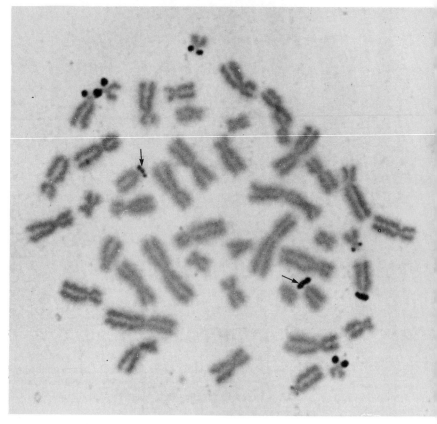

Figure 8.4 A normal human metaphase set stained with the 'AgAS' technique, which produces small black deposits of metallic silver at nucleolus organizers that were associated with a nucleolus in the previous interphase. In this picture, four D-group and four G-group chromosomes are stained with silver. At the top left-hand side of the picture a D-group and a G-group chromosome are associated end to end and share a large silver grain. Note how some chromatids carry a single silver grain, while others have two, making three grains per chromosome (arrows). Remember, there are altogether ten chromosomes that carry NORs in human, but only eight of these are stained in this preparation. (Reproduced with the permission of John Wiley & Sons, from Macgregor, H.C. and Varley. J.M., *Working with Animal Chromosomes,* 2nd edn, 1988.)

RINGS, CHRISTMAS TREES AND TRANSCRIPTION UNITS

There is little to be gained by looking at a nucleolus or a secondary constriction with an electron microscope, and at one time it seemed most unlikely that electron microscopy would contribute anything to this particular field of study. However, in the late 1960s the rediscovery of ribosomal gene amplification in oocytes and the development of techniques for handling oocyte nuclei and their giant lampbrush chromosomes completely changed that outlook. Oocyte nuclei had thousands of nucleoli. Each of these possessed its own small piece of rDNA that was presumed to be a replica of part or all of the chromosomal nucleolus organizer. Oocyte nuclei could be isolated manually, their membranes removed and their contents of chromosomes and nucleoli spilled out onto a slide for subsequent examination with a light microscope. Could not a technique be invented that would allow the isolation of these oocyte nucleoli and the dispersal of their substance so as to leave only the rDNA and its attached nascent transcripts of ribosomal RNA precursor molecules? Would it not then be possible to see these transcribing ribosomal genes with an electron microscope? Such a method was a product of the skill and ingenuity of Oscar Miller and his associates, then at Oak Ridge National Laboratory in Tennessee. It has since proved to be of exceptional value for the study of nucleolar function, RNA transcription and gene structure and evolution.

In 1964, two American investigators independently observed that the nucleoli of amphibian oocytes are not always round, solid structures, but they sometimes take the form of rings or beaded necklaces (Figure 8.5). The same phenomenon had been seen and reported some 20 years previously, but no-one had attached

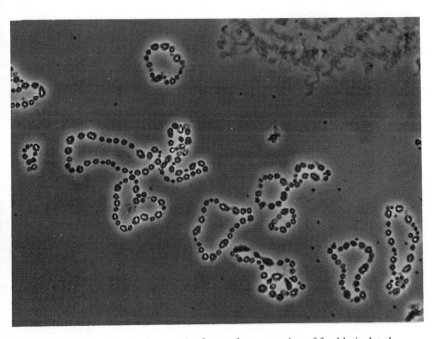

Figure 8.5 Phase-contrast micrograph of part of a preparation of freshly isolated chromosomes and nucleoli from a salamander oocyte. Some of the lampbrush chromosomes are visible in the top right-hand corner of the picture. The remainder of the objects in the picture are examples of the hundreds of 'beaded necklace' nucleoli that are found in these oocytes. Note how the beaded necklaces come in a range of sizes from the very small ring at centre left to the very long one next to it on the right.

much significance to it. These beaded necklace-type nucleoli could be broken into fragments by the enzyme deoxyribonuclease, thus confirming the presence of DNA in the nucleoli and demonstrating that the DNA component was in the

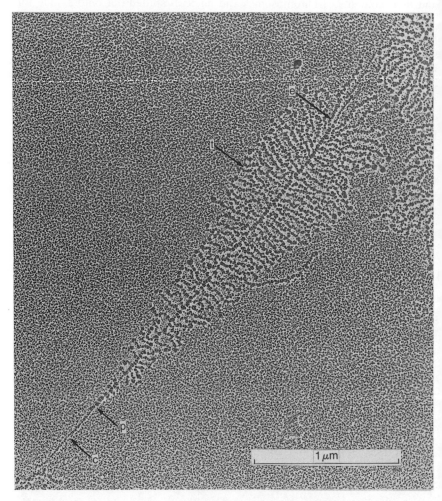

Figure 8.6 An electron micrograph of a single ribosomal gene transcription unit ('Christmas tree') prepared by the 'Miller spreading' technique involving the isolation of the chromosomes and nucleoli of an amphibian oocyte in pH 9 detergent water followed by centrifugation of the nucleolar DNA and its associated RNA polymerase and attached transcripts onto an electron microscope specimen support or grid. c, non-transcribing chromatin of the spacer region. p, region occupied by the first polymerase molecules at the very start of the transcription unit. t, closely packed RNA transcripts extending sideways from the DNA axis (the 'trunk' of the Christmas tree) each attached to the DNA by its polymerase. e, the end of the transcription unit. *Measure the length of this transcription unit. How does it compare with the length of the transcribed region of the gene shown in Figure 8.1?*

form of a continuous ring several tens or hundreds of micrometres long. Miller proceeded to isolate oocyte nucleoli in slightly alkaline water with a trace of detergent added, and he centrifuged them through an underlying layer of formalin and sucrose until they came to rest on an electron microscope specimen-supporting grid at the bottom of a specially constructed centrifuge chamber. The effect of the alkaline detergent water was to remove everything from the nucleolar DNA except the nascent RNA molecules and the polymerases by which their ends were attached to the DNA template. The formalin/sucrose acted as a histological fixative and a cushion to discourage sedimentation of debris from the oocyte nucleus. The result was remarkable and an example of it can be seen in Figure 8.6.

At this point it is important that we fully understand the relationship between a typical solid nucleolus, the beaded necklace nucleolus and the dispersed ring

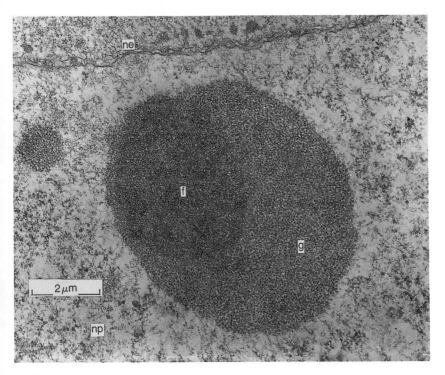

Figure 8.7 An electron micrograph of a thin section through a typical solid nucleolus in an amphibian oocyte. The nuclear envelope (ne) runs across the top of the picture. np, nucleoplasm. f, the fibrous core of the nucleolus which includes the nucleolar DNA and is the site of transcription of ribosomal RNA precursor. g, the granular cortex of the nucleolus which is a region occupied by ribosomal ribonucleoprotein (RNP) particles that will later be released from the nucleolus, exported to the cytoplasm and assembled into mature ribosomes.

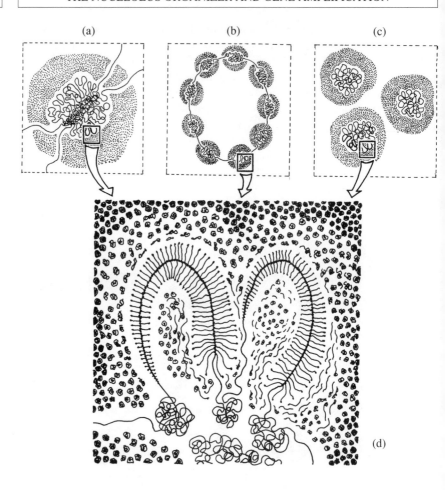

Figure 8.8 Schematic diagram to show the structural relationships between solid round and beaded necklace-type nucleoli such as are sometimes found in oocyte nuclei. (a) A schematic representation of a normal nucleolus that forms around the nucleolus organizer region on a chromosome, with the transcribing nucleolar DNA (= ribosomal genes) occupying much of the fibrous core of the nucleolus and surrounded by nascent ribosomal subunit particles in the granular cortex. (b) A beaded necklace nucleolus such as is found in some amphibian oocyte nuclei. Here each bead is essentially a 'mini-nucleolus' attached to its neighbours by an intervening portion of the continuous nucleolar DNA circle. (c) The situation in an oocyte nucleus where the nucleoli are not associated with a chromosome but have their own replicas of the nucleolar DNA structurally arranged in the same manner as in a normal somatic nucleolus. (d) A supposed arrangement of active transcription units around the periphery of the fibrous region in each type of nucleolus. None of the aspects of these diagrams are necessarily correct, but they are all based on currently available evidence and accepted views. Remember, beaded necklace nucleoli do not occur in situations where there is no amplification of ribosomal genes: they are based on the circular replicas of tandemly arranged ribosomal gene repeats that are the products of amplification by rolling circle replication.

of nucleolar DNA that we see in a Miller preparation. Figure 8.8 shows these relationships. Figure 8.9 gives the dimensions that characterize a Miller spread of a nucleolus from *Xenopus laevis*. The solid nucleolus consists of a granular cortex that is made up mainly of ribosomal subunits in various stages of completion. This encloses a more fibrous core that contains the DNA component and its attached RNA transcripts with associated proteins (Figure 8.7).

In the beaded necklace, the DNA component is extended into a ring, and at intervals along its length there are local accumulations of core and cortex material. The dispersed ring, as seen in a Miller preparation, takes the form of a thin strand of deoxyribonucleoprotein, parts of which have associated RNA transcripts. Regions that are clothed in transcripts are called **matrix units**. The matrix units are rather regular in size, and their transcripts increase in length from one end of the matrix unit to the other, reflecting the movement of RNA polymerase along the DNA template and the progressive elongation of the newly forming RNA molecules. Regions between the matrix units are called

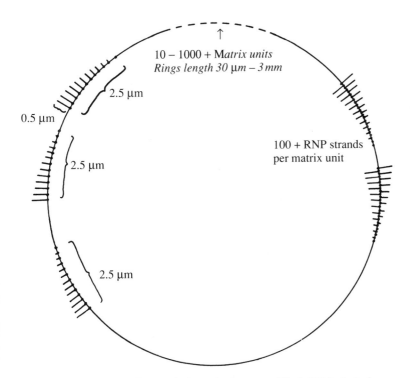

Figure 8.9 Arrangement of transcription units in an amplified rDNA circle from a *Xenopus laevis* oocyte. Note the variable length of the regions between transcription units (non-transcribed spacers, NTS). *How do the dimensions shown in this diagram compare with those seen in Figures 8.1 and 8.6?*

spacer segments and they are shown in the diagram as being of different lengths; this is an important feature, as we shall see later on.

UNDERSTANDING AMPLIFICATION

The nucleolus organizer does some very peculiar things, and I wish now to examine some of them carefully from the functional and evolutionary stand-points. Let us begin with gene amplification. In this regard, three matters relate to the general framework of this book. First, all those amphibians and fish that have been studied seem to amplify their ribosomal genes in their oocytes up to approximately the same level. Essentially, they all make the same amount of amplified rDNA, irrespective of the sizes of their genomes, the sizes of their nucleolus organizers, the number of chromosomal nucleolus organizers that they have or the means by which the amplification is accomplished. The last point refers particularly to an incredible situation in the tailed frog of the Cascade Mountains of western North America (*Ascaphus truei*), which amplifies its ribosomal genes by first making its oocytes octanucleate – a product of three rounds of oogonial mitosis without cell division – and then allows each of the eight nuclei to synthesize just one-eighth of the amount of amplified rDNA that is normally present in the single oocyte nucleus of related frogs (Figure 8.10). This quantitative regulation, notwithstanding the numbers of ribosomal genes that are available at the outset as potential primary templates, is all the more remarkable when we think that the regulatory mechanism must operate to shut down not just one replicating gene complex, but each and every one of thousands of secondary templates that are actively rolling off new rDNA to-wards the end of the process. It is a process that starts off slowly and then gathers momentum as more and more new rDNA is made and becomes available for subsequent template activity, then shuts down quite suddenly when the critical mass of rDNA has been produced.

It is worth thinking for a moment about a practical matter relating to rDNA amplification. We have already seen that at the end of amplification in *Xenopus*, two-thirds of the nuclear DNA is rDNA. Accordingly, *Xenopus* pachytene nuclei stained with the Feulgen reaction appear large and densely pink in colour, in contrast to the smaller diploid somatic cell nuclei that may lie nearby in a cytological preparation. In an animal with a larger genome the situation is rather different. The common crested newt (*Triturus cristatus*), for example, has a C-value of 22 pg. This means that its pachytene nuclei will have 4×22 pg; remember, they are diploid nuclei that have gone through the premeiotic S-phase to give each chromosome two chromatids. The *Triturus* nucleus makes the same amount of amplified rDNA as *Xenopus*. So it will have a total of 88 pg of chromosomal DNA plus a further 30 pg of amplified rDNA. In this instance, only about one-quarter of the total nuclear DNA is rDNA, so the pachytene nuclei at the end of the amplification process are not particularly conspicuous. It

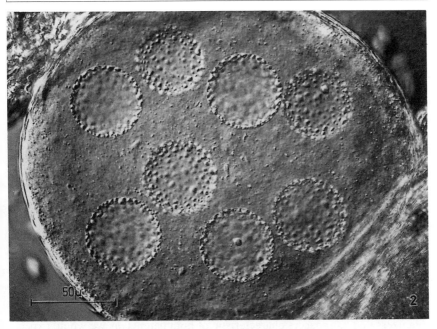

Figure 8.10 This remarkable picture shows an entire developing ovarian egg (oocyte) from the Pacific tailed frog, *Ascaphus truei*, photographed in a fresh and life-like condition with 'Nomarski interference contrast' optics. Nomarski microscopy has the unique advantage of providing a clear image of one focal plane through the middle of a thick object without the blurring effect of the out-of-focus material above and below the focal plane. Here we see the oocyte with eight large round nuclei. Each of these has a bubbly appearance owing to the presence of hundreds of round, solid nucleoli. This animal uniquely employs two strategies to amplify the ribosomal genes in its oocytes: it makes the oocyte eight-nucleate by preventing cell division in the last three rounds of oogonial mitosis, and then it amplifies its ribosomal DNA in each nucleus to one-eighth the level normally found in species whose oocytes have just one nucleus. (Reproduced with permission of Springer-Verlag from Macgregor, H.C. and Kezer, J., *Chromosoma (Berl.)* **29**, 189–206, 1970.)

is partly for this reason that rDNA amplification went unnoticed for so many years. It was described first in a species of *Bufo* that has a C-value intermediate between that of *Xenopus* and *Triturus*. It was not until *Xenopus* became the specific target for cytological studies of amplification that the phenomenon was convincingly demonstrated in Feulgen-stained cytological preparations.

The second point about amplification is closely allied to the first, but it deserves special mention. It is simply that if an organism has two nucleolus organizers at corresponding sites on homologous chromosomes, and we delete one of them to make a 'nucleolar heterozygote', then that organism nevertheless amplifies its rDNA in its oocytes to the same level as in the normal homozygote. This is to say, the level of amplification is independent of the numbers of chromosomal nucleolus organizers in the cell at the start of the process. Remember that we have already encountered a rather similar situation with regard to levels of polytenization of ribosomal genes in *Drosophila* that have different numbers of X chromosome nucleolus organizers.

The third and last point that I wish to make about amplification concerns the nature and specificity of the primary template. The question is simple and the manner in which it was answered is a good example of an experimental approach that required quite a wide knowledge of the biology of *Xenopus*. It is now known that the primary templates are certain of the repeat rDNA sequences in the chromosomal nucleolus organizers of the oogonium. The validity of this comment rests on the following observations. First, we know that the spacer nucleotide sequences of the rDNA of *X. laevis* are different from those of the rDNA of *X. mulleri*, such that the two rDNAs can quite easily be distinguished from one another. F1 hybrids between *X. laevis* and *X. mulleri* are fertile, and it can be argued that if gene amplification were initiated by a maternally inherited descendant from the amplified rDNA of the previous generation – the one serious alternative to initiation by rDNA of the chromosomal nucleolus organizers – then the amplified rDNA spacer would always be that of the mother species. In fact, hybrid females have rDNA in their chromosomal nucleolus organizers that has spacers characteristic of both parents, as we would expect, but the spacer of the amplified rDNA from the oocytes of the F1 hybrid females is always that of *X. laevis,* no matter whether the *laevis* parent was the mother or the father. This result effectively eliminates all possibility of the maternal inheritance of the primary template for rDNA amplification and it points to the chromosomal nucleolus organizers for this role.

Why, one must now ask, should the *laevis* template be 'dominant' in this respect? Are there species whose nucleolus organizers would override those of *laevis* in a hybrid situation, or that would be subordinate to those of *borealis*? Do the two nucleolus organizers of a hybrid behave competitively in other respects? Does this kind of 'dominance' or competition operate at other levels? Does it, for example, come into the same category as the selective transcription of mouse nucleolus organizers in mouse–human hybrid cells? The specific answers to these questions are less important here than the questions themselves. The questions are important because they show us how this kind of science progresses and how the minds of experimental biologists work.

Ribosomal DNA amplification is not accomplished by semiconservative replication of the kind that happens during normal S-phase replication of chromosomes. It proceeds by a 'rolling circle' mechanism that is able to generate closed circles of DNA of widely variable length (Figure 8.11), thus accounting for the different lengths of beaded necklace nucleoli in oocyte nuclei (Figure 8.5). It is worth noting at this point that there are several other situations in which specific gene complexes are amplified as a component of some intensified process in a specialized cell type. The best-known example is the genes for chorion (egg case) proteins of the *Drosophila* egg. These genes

become amplified in the follicle cells that surround the developing egg immediately before the start of synthesis of chorion proteins and subsequent laying
down of the egg case. Here the mechanism of amplification *is* semiconservative,
but it is adapted to ensure that all the amplified copies of the genes remain firmly
integrated into the chromosome, rather than dispersing as independent
'episomes' as we have seen with rDNA. The DNA puffs that were mentioned in
Chapter 7 are another case of amplification along the lines of the chorion protein
system, although it has not been fully investigated in this regard.

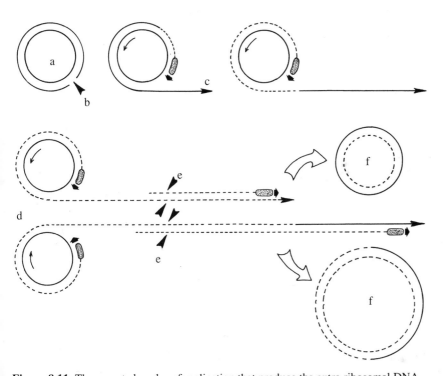

Figure 8.11 The repeated cycles of replication that produce the extra ribosomal DNA
copies in oocyte nuclei have been shown to occur by the rolling circle mechanism, a
process known to be involved in the replication of some viral genomes. The mechanism
requires that a segment of the chromosomal ribosomal gene complex be released from
the chromosome as a covalently closed circular molecule (a). Replication of the circle
starts at a single-strand nick in the circle (b) from which the free 5' end unwinds and
extends as a tail to the side of the circle (c). As the tail extends, the circle rolls in the
direction of the arrow, and replication begins on both of the single-strand regions
exposed by the unwinding. On the circle, new nucleotides are added continuously to the
exposed 3' end using the un-nicked circle of DNA as an endless template. On the tail,
new nucleotides are added to the complementary copy that extends continuously as the
tail unwinds (d). At apparently random intervals the tail is clipped off (e) and the free
ends of the linear double-stranded molecules are covalently sealed together to form
more closed circles of variable length (f).

VARIATION AND EVOLUTION

The next feature of nucleolus organizers that deserves attention is the variability in numbers of ribosomal repeats and the sizes of nucleolus organizers within species. Until about 1970 it was generally assumed that nucleolus organizers are of about the same size in all individuals belonging to the same species. Early studies of the ribosomal genes of *Xenopus*, for example, led to the general concept that all individuals have two nucleolus organizer chromosomes per cell, each of which carries approximately 450 copies of the DNA sequences associated with the synthesis of ribosomal RNA. Measurements on other animals produced values that were supposed to be characteristic of the species. The ribosomal genes are important, so it was natural to expect that they would be rather well conserved in sequence and in repetitive frequency, at least within species and probably also between related species. Neither of these expectations proved to be correct.

In the first place, different species have widely different average numbers of ribosomal repeats per diploid cell. In general, the larger the genome, the more ribosomal repeats it contains, although this relationship is by no means linear. *Tetrahymena*, a ciliated protozoon, has just 20 ribosomal repeats. *Drosophila melanogaster* has about 150. *Xenopus* and *Homo* have about 400. Certain salamanders belonging to the genus *Plethodon*, with genomes about ten times as large as those of *Homo* or *Xenopus*, have between 500 and 5000 ribosomal repeats.

Then there is the question of variability *within* one species. This can be quite striking and highly significant. The first clear indication that different individuals within a species might have different numbers or ribosomal gene repeats came from studies of 22 males and 38 females of the toad *Bufo marinus* in 1969. Here it was shown that different individuals have nucleolar secondary constrictions of different lengths, that a single individual can be heterozygous for the sizes of its homologous nucleolus organizers, and that there is a two-fold range of variability from animal to animal within the species with respect to the proportion of the entire genome that is complementary in DNA sequence to 28S and 18S rRNA.

A little later, certain mutants of *Xenopus laevis* were characterized as having different numbers of ribosomal repeats, with variability from one animal to another, and it was estimated that the minimum number of ribosomal repeats necessary for normal growth and development of a *Xenopus* is between one-quarter and one-half of the average number that had hitherto been regarded as typical for the species.

If we look at a natural population of animals, all belonging to the same species, the true range of variability with regard to rDNA becomes apparent. The North American red-backed salamander, *Plethodon cinereus,* offers a good example because it is common, it exists in dense natural populations that are easy to sample and it has large chromosomes that are exceptionally good for cytological studies. In this species there is a wide variability in the numbers of rDNA repeats per diploid cell from one individual to another. The range of this variability in one population is quite different from that in another. A population from a small wildlife reservation in the state of New Jersey in the USA showed a basic level of variability over a 7.5-fold range, whereas another from Connecticut showed variability over a 2.5-fold range. Most of these animals were strongly heterozygous with respect to the sizes of their nucleolus organizers (numbers of rDNA repeats per nucleolus organizer), so that in any one animal it was likely that one nucleolus organizer carried at least twice as many rDNA copies as its partner on the homologous chromosome. Consider, for example, two animals that showed 630 and 4800 rDNA repeats (actual observed values) per diploid cell. Suppose that in each of these one nucleolus chromosome carried twice as many rDNA repeats as the other. Accordingly, the smaller nucleolus organizer in the first animal would have 210 rDNA repeats and the larger one in the second animal would have 3200, representing an actual range in nucleolus organizer size of at least 15-fold.

The questions that spring to mind at this point are really quite fundamental. How does this variability in the sizes of nucleolus organizers come about? Does each population within a species have its own peculiar range of variability with respect to this gene complex, and if so is there some selective advantage associated with that range? Can the range of variability tell us anything about the ages or evolutionary histories of populations? Does this kind of variability exist with other repetitive genes, such as, for example, the genes for histones or the other ribosomal molecule, 5S RNA?

So far we have concerned ourselves with quantitative variability. What about the DNA sequences of ribosomal genes and their spacers? Once again, *Xenopus* is a good starting point for this topic. In *X. laevis* and *X. mulleri* the sequences that actually code for 18S and 28S rRNA are almost identical. The small transcribed spacer (Figure 8.1), located between the 18S and 28S genes is of the same length in both species but different in sequence. The large spacer, most of which is not transcribed, is of the same average length in both species but quite different in sequence: 73% of the bases (nucleotides) in the *laevis* spacer are either G or C, whereas only 69% of the bases in the *mulleri* spacer are G or C. The two spacers have been estimated to differ by at least 1000 of their 7000 base pairs. There are variations in the lengths of the large spacers, and these reflect variable numbers of internal subrepeats within this region.

How well are the actual ribosomal DNA sequences, the coding sequences, conserved? In fact, remarkable poorly when we consider that they are the basis of an end product, the ribosome, that must be one of the oldest and longest established components of all living cells. True, the ribosomal genes are the same in closely related species like *X. laevis* and *X. mulleri*, but if we look further afield then the picture changes. The 28S molecule of *Drosophila hydei*, for example, has a molecular weight of 1.43×10^6 daltons. That of *X. laevis* is 1.5×10^6 daltons. The 28S molecule of *Xenopus* has stretches comprising a total of only 360 out of 4300 nucleotides that are common to both mouse and *Drosophila*. Likewise, for the 18S molecule the common regions amount to no more that 350 nucleotides out of a total of around 2000. The general level of similarity is naturally higher the closer the organisms are in the phylogenetic sense and, as with *Xenopus*, the ribosomal sequences of species within the same genus are closely similar.

It seems, then, that we are dealing with five different levels of evolution or change in nucleolus organizers and ribosomal genes. First, we have ribosomal gene-coding sequence divergence, proceeding at a slow pace and probably quite strictly controlled by natural selection in certain critical parts of the coding sequences. Second, we have a much more rapid spacer sequence divergence, signifying that the actual DNA sequence of the spacer region is of much less critical importance than that of the coding sequences. Third, we have variability in the repetition frequency of the entire spacer/gene unit, regulated at the bottom end by the need to maintain a certain minimal number of repeats per diploid cell. Fourth, we have variability in the number of internal subrepeats within the non-transcribed spacer regions, making spacers of different lengths. Fifth, we have random assortment in meiosis of nucleolus organizers of different sizes and on different chromosomes. It is in these senses that the nucleolus organizer presents us with a superb model system for investigating and understanding evolution at the gene and chromosome level in eukaryotic cells.

THE ACTIVITY OF NUCLEOLUS ORGANIZERS

One last question: does variability in the sizes of nucleolus organizers have anything to do with nucleolus organizer activity? We have seen that in the same cell some nucleolus organizers are 'active' and others are not. The man–mouse hybrid cells and the *X. laevis*/*X. mulleri* hybrid animals offer good examples with respect to transcription and amplification respectively. Is differential activity a general phenomenon and, if so, then is the activity of a nucleolus organizer a function of its size, or perhaps its spacer sequence, or what?

A useful experimental approach to this kind of question involves cells in tissue culture. For example, there happen to exist two cell culture lines from *Xenopus laevis* that have one and three nucleolus organizers per cell. We should take care at this point to note that these are cultured and transformed cells that are no longer diploid. Their nucleolus organizers are therefore on different chromosomes and they do not have counterparts on homologous chromosomes. Besides these lines it is possible to set up new or primary culture lines that have karyotypes closely similar, if not identical, to that of the live animal. Primary lines have two nucleolus organizers in corresponding positions on homologous chromosomes. Cells in tissue culture can, of course, be grown and supplied with radioactive precursors in a controlled fashion, and they are generally good for quantitative studies. We therefore have the opportunity to compare numbers of nucleolus organizers, located by secondary constrictions and/or *in situ* hybridization, with numbers of ribosomal genes (rDNA repeats), determined by quantitative nucleic acid hybridization, with activity of nucleolus organizers, assessed by silver staining and rates of incorporation of radiolabelled RNA precursors.

The results of this kind of experimental approach applied to the *Xenopus* cell culture system allow us to draw some very firm negative conclusions. The rate of synthesis of ribosomal RNA, which is a clear measure of the activity of a nucleolus organizer, is not a simple function of the size of the nucleolus organizer, the number of ribosomal gene repeats in the cell or the number of nucleolus organizers in the cell. Nor is it a function of the competitive superiority and selective transcription of the largest organizer in the cell.

In this regard, the human system is very interesting indeed. Normal men and women, just like amphibians and fruit flies, vary from one to another with respect to the number of rDNA repeats per diploid cell. The range probably extends between 250 and 500. All normal people have ten nucleolus organizers (five pairs) per diploid cell. They are located on the short arms of each of the D- and G-group chromosomes, numbers 13, 14, 15, 21 and 22. In no normal person do we find all ten nucleolus organizers staining with silver. The number of silver-staining sites varies from 3 to 7 with a mode of around 5. This is to say that, as a rule, only half the nucleolus organizer sites in *Homo* are synthetically active.

The plot thickens when we consider abnormal human karyotypes. Trisomy (three representatives) for chromosome 21 produces Down's syndrome. Chromosome 21 is a nucleolus chromosome. In patients suffering from Down's syndrome, nine, ten or all 11 of the nucleolus organizer sites stain with silver. The same applies to people with a chromosomal rearrangement in which the short arm of chromosome 9 is translocated to the short arm of chromosome 13, again a nucleolus chromosome, and this applies whether the translocation is **balanced** (only two short arms of chromosome 9 present, one of them translocated) or **unbalanced** (three short arms of chromosome 9 present, two normal chromosomes 9 and one translocated short arm). The same does not apply to translocations that do not involve nucleolus chromosomes. For example, people with trisomy 18 show the normal mode of silver-staining nucleolus organizers.

Data on human systems have led to the following conclusions. Activity of a nucleolus organizer is not a function of its size. Nor is it a function of its spacer sequences, since both members of a homologous pair of nucleolus organizers do not necessarily stain with silver. Activity is a heritable and individual specific character in the sense that each person shows a consistent life-long pattern of silver staining of his or her nucleolus organizers sites; but, you will be quick to ask, what happens if a male and a female each contribute nucleolus-active chromosomes to one of their offspring to make a total of more than the mode of 5? Does the mysterious 'dominance' factor then come into play such that some organizers that were active in the parent are suppressed in the child? Whatever its basis, the regulation of activity is quite finely balanced and can be switched by introducing structural rearrangements that involve chromosome breaks or rejoins in the vicinity of one or other of the nucleolus organizers.

RECOMMENDED FURTHER READING

Fischer, D., Weisenberger, D. and Scheer, U. (1991) Assigning functions to nucleolar structures. *Chromosoma (Berl.)* **101**, 133–140.

Hernandez-Verdun, D. (1991) The nucleolus today. *J. Cell Science* **99**, 465–471.

Jordan, E.G. (1991) Interpreting nucleolar structure: where are the transcribing genes? *J. Cell Science* **98**, 437–442.

Jordan, E.G. and Cullis, C.A. (Eds.) (1982) *The Nucleolus*. Society for Experimental Biology Seminar Series 15. The University Press, Cambridge.

Macgregor, H.C. (1972) The nucleolus and its genes in amphibian oogenesis. *Biol. Rev.* **47**, 177–210.

Macgregor, H.C. (1982) Ways of amplifying ribosomal genes. In *The Nucleolus*. E. G. Jordan and C.A. Cullis (Eds.). The University Press, Cambridge, pp. 129–152.

Macgregor, H.C. and Kezer, J. (1970) Gene amplification in oocytes with 8 germinal vesicles from the tailed frog *Ascaphus truei* Stejneger. *Chromosoma (Berl.)* **29**, 189–206.

Lampbrush chromosomes | 9

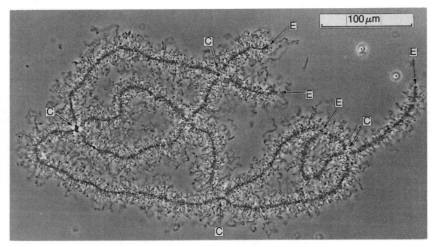

Figure 9.1 A phase-contrast micrograph of a freshly isolated lampbrush bivalent from the North American newt *Notophthalmus viridescens*. Arrows marked E indicate the four ends of the bivalent. Arrows marked C indicate the positions of chiasmata.

HISTORY AND TECHNOLOGY

Lampbrush chromosomes are something special for the cytologist. On the one hand they are big and they look quite different from any other form of chromosome. On the other hand, they have offered a successful medium through which it has been possible to draw valid conclusions at the molecular level on the basis of observations and experiments carried out at the level of the light microscope. People have not only looked at lampbrush chromosomes very carefully indeed, they have also been able to do things to them and actually watch the consequences as they are happening. Of course, the total volume of published research that has been carried out on polytene chromosomes from organisms such as *Drosophila* and *Chironomus* far exceeds that on lampbrush chromosomes. Nevertheless, lampbrush chromosomes have proved uniquely valuable in two respects. First, they are transitory structures that exist during an extended diplotene of the first meiotic division. The chromosomes go from a compact telophase form at the end of the last oogonial mitosis, become lampbrushy and then contract again to form perfectly normal first meiotic metaphase bivalents. Secondly, their most conspicuous feature is widespread RNA transcription from hundreds, and in some cases thousands, of transcription units that are arranged at short intervals along the lengths of all the chromosomes. In these senses it has been possible to exploit lampbrush chromosomes in the study of chromosome organization and gene expression during meiotic prophase, and in studies of the molecular and supramolecular morphology of RNA transcription.

Lampbrush chromosomes were first seen in sections of salamander oocytes by Flemming in 1882. Ten years later they were described in the oocytes of a dogfish by Ruckert (1892). The name lampbrush comes from Ruckert, who likened the objects to a nineteenth-century lampbrush, equivalent to the twentieth-century test-tube brush.

The lampbrush type of chromosome is now known to be characteristic of growing oocytes in the ovaries of almost animals with the exception of mammals and certain insects. The chromosomes are greatly elongated diplotene bivalents, sometimes reaching lengths of a millimetre or more.

Lampbrushes are exceedingly delicate structures and no further progress beyond the pioneer studies of Flemming and Ruckert was possible until a technique could be devised for dissecting them out of their nuclei and examining them in a life-like condition, separated from the remainder of the nuclear contents. This is not as difficult as it might sound. The largest lampbrush chromosomes are to be found in growing oocytes of newts and salamanders. We have already seen that these animals have big genomes, big chromosomes and big cells, so it is scarcely surprising that they have good lampbrushes.

The best oocytes for lampbrush studies are the ones that make up the bulk of the ovary of a healthy adult female at the time of year when the eggs are actively growing in preparation for ovulation in the following spring. They are about 1 mm in diameter. They have nuclei that are between a third and a half a millimetre in diameter, big enough to see with the naked eye. These nuclei are really not hard to isolate by hand and it is not much more difficult to remove their nuclear envelopes and spill out their chromosomes. Such a technique was introduced by Dr Joseph Gall in 1954, working in the University of Minnesota. The oocyte is submerged in a suitable saline solution, it is punctured with a needle, the nucleus is gently squeezed out of the hole, picked up in a Pasteur pipette and transferred to fresh saline in a chamber constructed by boring a hole through a microscope slide and then sealing a coverslip across the bottom of the hole with wax. The nuclear envelope is then removed and the nuclear contents, including the lampbrush chromosomes, come to lie flat, and hopefully unbroken and well displayed, on the bottom of the chamber (Figures 9.2 and 9.3). Then, by using a phase-contrast microscope with an inverted optical system, the chromosomes can be examined in a fresh and unfixed condition with the highest resolution and magnification obtainable with a light microscope.

The reason for using an inverted optical system is simple. If the chromosomes are dissected onto a slide in a drop of saline and then covered with a coverslip and looked at with a normal microscope, the very act of placing a coverslip on top of them, a combination of movement, turbulence and surface tension, completely destroys them. So we have to dissect them into a flat-bottomed chamber. When they are lying at the bottom of the chamber they remain undisturbed when we place a coverslip over the top of the chamber. If we then try to look at them from the top with an ordinary microscope, we are looking through the entire depth of the chamber, which will be a distance equal to the combined thickness of the microscope slide and the top coverslip. That distance

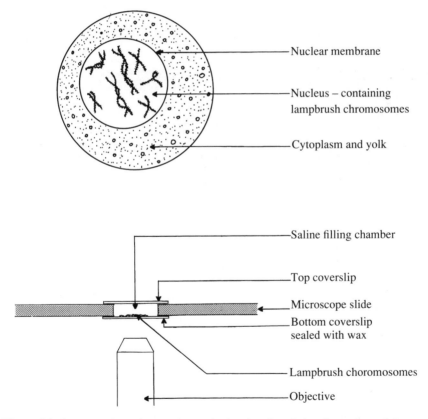

Figure 9.2 An oocyte (growing ovarian egg), showing the relative dimensions of the egg, its nucleus and its lampbrush chromosomes, and the system for visualizing the chromosomes using an inverted microscope and a chamber constructed from a slide with a hole bored through it.

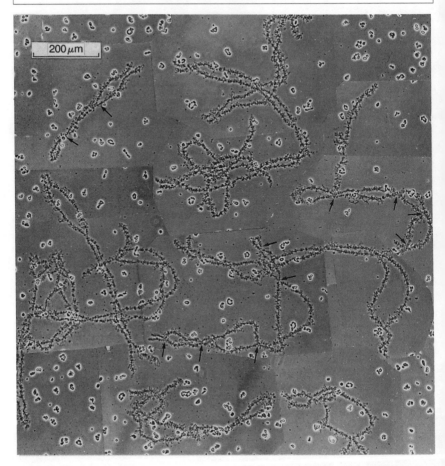

Figure 9.3 A composite of 15 phase-contrast micrographs covering the entire area of over 1 mm² occupied by a full set of 14 lampbrush bivalents freshly isolated from an oocyte of the plethodontid salamander *Plethodon cinereus*. The numerous small, bright dots and rings are the many nucleoli that are the products of gene amplification in the early stages of oogenesis (see Chapter 8). Note how each lampbrush bivalent consists of two half-bivalents stitched together at several points along their lengths by chiasmata (arrows indicate chiasmata on three of the bivalents). The scale bar in this collage represents one-fifth of a millimetre (200 μm). *How long is the longest bivalent?* (Picture reproduced with the kind permission of Dr Marcela Vlad of the University of Warwick.)

is greater than the working distance of most high-power microscope objective lenses. It will therefore be impossible to bring the chromosomes into focus. If, on the other hand, we look at them through the bottom of the chamber then we are looking through a distance equal only to the thickness of the coverslip that forms the floor of the chamber, well within the working distance of all objectives (Figure 9.2).

BASIC ORGANIZATION

The most important factor to keep in mind in relation to the structure of a lampbrush chromosome is that it is **a meiotic half-bivalent**. This means that it must consist of two chromatids. The entire lampbrush bivalent, of course, will have a total of four chromatids. The chromosome appears as a row of granules of deoxyribonucleoprotein (DNP), the chromomeres, connected by an exceedingly thin thread of the same material (Figure 9.4). Chromomeres are 0.25–2 μm in diameter and spaced 1–2 μm centre to centre along the chromosome.

Each chromomere has two, or some multiple of two, loops associated with it. The loops have a thin axis of DNP surrounded by a loose matrix of ribonucleoprotein (RNP). The loops are variable in length, and during the period of oogenesis when they are maximally developed they extend from 5 to 50 μm laterally from the chromosome axis, which means that the longest loops in such a case would be 100 μm long. The loops are also variable in appearance. Loops of the same appearance always occur at the same locus on the same chromosome, from one animal to another within a species. Accordingly, some loops with particularly distinctive appearances can be used reliably for chromosome identification. Very importantly, 'sister' loops, arising from the same chromomere, have the same appearance and usually, though not always, are of the same length (Figure 9.4).

In the European crested newt (*Triturus cristatus*) or the North American newt (*Notophthalmus viridescens*), the two animals on which most lampbrush studies

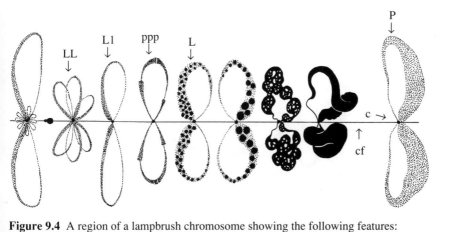

Figure 9.4 A region of a lampbrush chromosome showing the following features: (1) interchromomeric axial fibre (cf) connecting small compact chromomeres (c); (2) chromomeres bearing pairs (L) or multiples of pairs (LL) of lateral loops; (3) loops of quite widely different morphologies; (4) sister loops of the same or different (Ll) lengths; (5) chromomeres without loops; (6) polarization of thickness along loops (P); (7) loops consisting of a single unit of polarization (P); (8) loops consisting of several tandem units of polarization with the same or different directions of polarities (ppp).

have been carried out, the chromosomes are quite short and contracted at the end of pachytene in the female. They then become greatly extended and assume the lampbrush form and they remain like that for several months. As the oocyte nears maturity, the loops and chromosomes quite suddenly become shorter, the chromomeres become larger, and eventually the chromosomes come to look like normal condensed diplotene bivalents. The general pattern of events is one of extension followed by retraction of the lampbrush loops and a clear inverse relationship between loop length and chromomere size. The longer the loop, the smaller the chromomere, and vice versa.

Most lateral loops have an asymmetrical form. They are thin at one end of insertion into the chromomere and become progressively thicker towards the other end (Figure 9.4). When one stretches a lampbrush chromosome, either deliberately or accidentally, breaks first happen transversely across the chromomeres in such a way that the resulting gaps are spanned by the loops that are associated with the chromomeres (Figure 9.5). Breaks of this kind are referred to

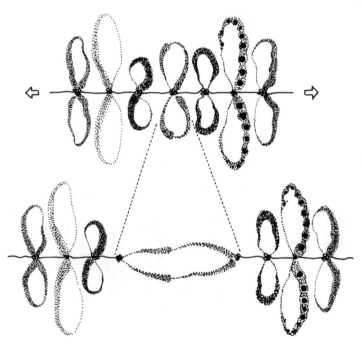

Figure 9.5 The formation of a double bridge break brought about by characteristic fracture of a stretched lampbrush chromosome across a line of weakness in a chromomere, such that the loops associated with that chromomere come to span the gap created by the fracture: an important observation providing additional evidence of the two-chromatid structure of the chromosome, with chromatids closely fused into one fibre in the interchromomeric regions but separated from one another in the regions of the loops.

in the lampbrush literature as 'double bridge breaks'. Clearly, double bridge formation indicates that there must be a line of weakness separating the two halves of a chromomere and, more importantly, it indicates a structural continuity between the main axis of the chromosome – the interchromomeric fibre – and the axes of the loops, in which connection we should remember again that these chromosomes are two-chromatid meiotic half-bivalents.

The last two basic points that need to be made about lampbrush chromosomes and their loops are very important ones indeed, since they serve as foci for the major questions that have been intensively investigated from 1954 right through to the present day. First, the loops are sites of active RNA synthesis, and in the vast majority of cases RNA is being transcribed simultaneously all along the length of the loop. In the crested newt there are more than 20 000 RNA-synthesizing loops per oocyte. Secondly, it has become clear that within a species particular loops may be present or absent in homozygous or heterozygous combinations, and if one examines the frequency of combinations within and between bivalents with respect to presence or absence of particular loops, then we find that these characters assort and recombine like pairs of Mendelian alleles. In other words, there appears to be an element of genetic unity in a loop–chromomere complex.

All the facts that I have given so far were known by 1960, just six years after Gall first inverted his phase-contrast microscope and made the detailed study of

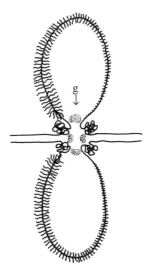

Figure 9.6 Currently accepted model of lampbrush organization, emphasizing the two-chromatid structure of the chromosome, a chromomere that must comprise four parts held together by some kinds of 'glue' proteins (g) without intimate intertwining of the chromatin fibres in each part, and sister loops that have similar, though not necessarily identical, lengths, and the same directions of thickness polarity.

lampbrushes possible. What then emerged was a model and two interesting and related hypotheses.

The model depicted a lampbrush chromosome as two DNA duplexes running alongside one another in the interchromomeric fibre, compacted into chromomeres at intervals and extending laterally from a point within each chromomere to form loops where RNA transcription takes place (Figure 9.6). Each duplex represents one chromatid.

TWO FAMOUS HYPOTHESES

The two hypotheses that followed this model have not stood the test of time or experiment, but both have stimulated a lot of thought and research from which there has been a remarkable spin-off of truth and understanding. In this sense we are reminded of the encouraging fact that a hypothesis does not have to be correct to be useful.

In the sequence in which they were developed, the first hypothesis took account of:

- the asymmetric shape of the loops;
- the inverse relationship between chromomere size and loop length;
- RNA synthesis on the loops.

It was referred to as the **spinning out and retraction hypothesis**. It said that during the lampbrush phase of oogenesis all DNA is progressively spun out from one side of the chromomere and the loop extends to become longer and longer. Subsequently, and at times simultaneously, loop axis DNA is retracted back into the other side of the chromomere, at which point it ceases to support RNA transcription. According to this hypothesis, all chromosomal DNA is likely to be involved in transcription at some time during the lampbrush phase. Loop asymmetry is accounted for by supposing that the portion of the loop that has been involved in transcription for the longest time will have the most material associated with it, and that will be the thick end of the loop, whereas the portion of the loop that is newly released from the chromomere in the spinning out process will have the least material associated with it, and that will be the thin end of the loop. This hypothesis, as we shall soon see, was later proved wrong.

The second hypothesis, which in modern terms was much more fundamental, made the point that the spinning out and retraction explanation for loop asymmetry included the assumption that there was no genetic diversity within an individual lampbrush loop–chromomere complex, and that the information carried in one of these complexes (an average one of which contains as much DNA as in the entire genome of the common colon bacillus *Escherichia coli*) is serially repeated along the entire length of the DNA that is located in the loop and its half-chromomere. As we shall see later, this assumption was unnecessary. At the time it was believed that the notion of serially repeated DNA sequences was incompatible with the fact that many phenotypically expressed mutations resulted from changes in only a few nucleotides, and it was hard to see how a mutation could possibly be expressed if it were not simultaneously imprinted in all copies of a repetitive gene complex. To overcome this dilemma, the then famous master/slave hypothesis was produced, in which the master sequence imprinted itself on all the slaves once per generation, and what better time to do it than during meiotic prophase.

These wonderful hypotheses emerged at what has to have been the most exciting and progressive era of chromosome research of the century. People talked about them, worried about them, designed experiments to test and extend them, tried to apply them to other situations, built models around them, and then finally demolished them.

The instruments of demolition were one straightforward observation and one simple series of experiments, both products of the new technologies of the 1970s. The observation came with the application of the Miller spreading technique in the study of amplified nucleolar DNA in oocyte nuclei (see Chapter 8). The experiments involved the technique of *in situ* nucleic acid hybridization.

LAMPBRUSHES UNDER THE ELECTRON MICROSCOPE

The same nuclei as were used for Miller spreads of nucleolar DNA also, of course, had lampbrush chromosomes in them. The lampbrush loops that appeared in these spreads looked just like very long transcription units. These transcription units, like the ones formed by ribosomal DNA, consisted of a thin DNA axis with RNA polymerase molecules lined up and closely packed along its entire length (Figure 9.7). Each polymerase molecule carried a strand of RNP. At one end of the transcription unit the RNP strands were short. At the other end they were much longer and they showed a smooth gradient in size from the one end to the other. In essence, the entire transcription unit was polarized, asymmetric, in the same sense as a loop, as seen with the light microscope, is asymmetric. The DNA axis before the start of the transcription unit and beyond its end showed the structure that would be expected of non-transcribing chromatin. The average lengths of chromosomal transcription units were about the same as the average lengths of loops as seen and measured with light microscopy.

At the time these observations were first made, it was well known that RNA transcription from a DNA template involved attachment of RNA polymerase to the DNA and movement of the polymerase along the template, feeding off nascent RNA as it went. The farther the polymerase travelled, the longer the piece of nascent RNA attached to the template by the polymerase. This was the mechanism that was held to account for the transcription of ribosomal RNA from the amplified rDNA in those same oocyte nuclei that had lampbrush

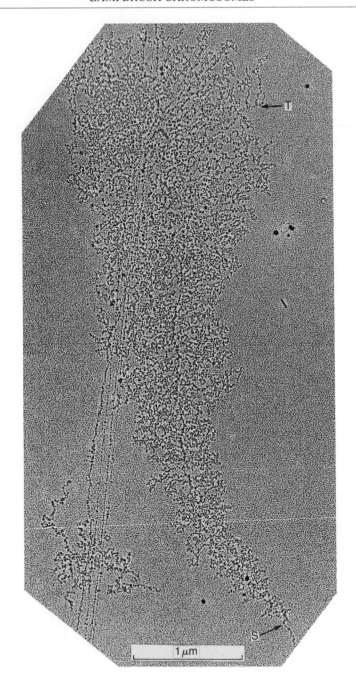

chromosomes. This mechanism alone was sufficient to account for the asymmetry of lampbrush loops, and in relation to loops it was inconsistent with the idea of a moving loop axis. Surely it was the polymerase that moved on a stationary axis, not the loop axis itself. Accordingly, it seemed simpler to suppose that a loop formed by an initial 'spinning out' process, probably powered by the continuing attachment of more and more polymerases to a specific region of the chromomeric DNA. It then remained and was transcribed as a permanent structure throughout the lampbrush phase. Towards the end of the lampbrush phase, transcriptive activity would decline, polymerases would detach from loop axes, and loops would regress and disappear. In the sense of the old hypothesis, there was no continuous spinning and retraction. The vast majority of the chromomeric DNA was never transcribed and a loop represented a short specific part of the DNA in a loop–chromomere complex.

IN SITU HYBRIDIZATION

In situ nucleic acid hybridization has already been mentioned in Chapter 8 as a means of locating specific gene sequences on chromosomes. It has a very special usefulness in relation to the study of lampbrush chromosomes. Let us suppose that each loop represents 'a gene'. The RNA that makes up the loop matrix, the attached nascent transcripts, will all be or include transcripts of that 'gene'. In effect, the loop represents a very large object, consisting of hundreds of RNA copies of the gene, all clustered at one position on the chromosome set. If we can isolate and purify the DNA of that gene and label it with a radioisotope, then it will be easy to make it single-stranded, and bind it specifically to the complementary single-stranded RNA attached to the lampbrush loop. The technique is known as **DNA/RNA transcript *in situ* hybridization (DR/ISH)**. How did it help to demolish the spinning out and retraction and the master/slave hypotheses?

The end product of an experiment involving DR/ISH is an autoradiograph showing one or more pairs of loops with silver grains distributed along their lengths (Figure 9.8). Two particular observations were of crucial importance. First, it is not uncommon in DR/ISH experiments to find loops that are labelled

---

Figure 9.7 An electron micrograph of a single transcription unit from a lampbrush loop prepared by the Miller spreading technique in which most of the chromosomal protein is removed by treatment in pH 9 detergent water, leaving only the DNA axis and its associated RNA polymerase and nascent RNA transcripts. The unit starts at the bottom right-hand corner of the picture (S). Its overall length is 5.2 μm. The axis of the unit is studded with RNA polymerase molecules. Note the complex form of the laterally projecting transcripts, remembering that the transcript at T must actually be as long as the DNA from which it was transcribed, which is the distance from S to T. (Reproduced with permission from Macgregor, H.C., *Heredity* **44**, 3–35, 1980.)

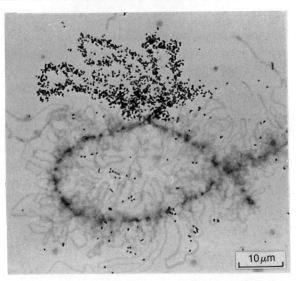

Figure 9.8 Autoradiograph of part of a lampbrush chromosome from the European crested newt following *in situ* nucleic acid hybridization of a single gene probe to newly transcribed RNA associated with lampbrush loops. The picture shows two pairs of large loops specifically labelled with silver grains, indicating that RNA transcripts containing sequences complementary to the radioactive probe sequence were distributed all along the lengths of these loops and sequences identical to the probe were represented in the DNA axis of the loop.

over only part of their lengths (Figure 9.9). This can be interpreted as evidence that the DNA sequence of a loop axis can and does change from place to place along the length of the loop. Evidence of that kind is incompatible with the master/slave hypothesis, which is based on the principle that entire loop–chromomere complexes consist of multiple tandem repeats of the same sequence. Second, wherever we find partially labelled loops in a DR/ISH experiment, it is usual to find the same partially labelled loops, with precisely the same pattern of labelling, in every oocyte over quite a wide span of size and stage. Such evidence cannot be reconciled with a dynamic spinning out and retraction of loop axis throughout the entire lampbrush phase.

The two classic hypotheses were replaced in the late 1970s and early 1980s by a much clearer understanding of the mechanism of action of lampbrush chromosomes and their loops. One of the first and most surprising discoveries that emerged from DR/ISH experiments was that highly repeated short DNA sequences, commonly referred to as 'satellite' DNA, sequences that could not possibly serve as a basis for transcription and translation into functional polypeptides, were abundantly transcribed on lampbrush loops along with more complex sequences that were definitely translated into functional proteins.

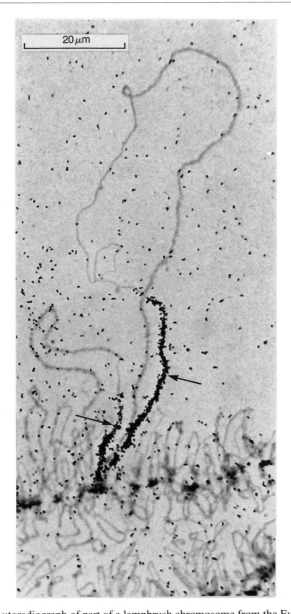

Figure 9.9 Autoradiograph of part of a lampbrush chromosome from the European crested newt following *in situ* hybridization of a single gene probe (in this case, part of the gene complex that codes for ribosomal RNA) to RNA transcripts associated with lampbrush loops. The two longest loops in this photograph are labelled part way along their lengths from one end. They are sister loops, arising from the same chromomere, yet they are of different lengths and their labelled segments are correspondingly different in length. (Micrograph reproduced with the kind permission of Dr G.T. Morgan from Morgan, G.T., Macgregor, H.C. and Colman, A., *Chromosoma (Berl.)* **80**, 309–330, 1980.)

THE READ-THROUGH HYPOTHESIS

Following this discovery, a strong new hypothesis quickly evolved. At the thin base of each loop or the start of each transcription unit there is a promoter site, a start signal, for a functional gene sequence. RNA polymerase molecules attach to this site and proceed to move along the DNA, transcribing the sense strand of the gene and generating messenger RNA molecules that remain attached to the polymerase (Figure 9.10). In the lampbrush environment there are no definitive stop signals for transcription, so the polymerases continue to transcribe past the end of the functional gene and into whatever DNA sequences lie 'downstream' of the gene. The consequences are very long transcription units, very long transcripts, mixing of gene transcripts with nonsense transcripts in high

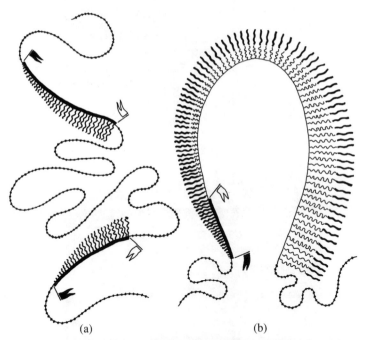

(a)　　　　　　　　　　　　　　　　(b)

Figure 9.10 (a) depicts a pattern of transcription that may be expected in normal diploid somatic cells, where genes are spaced out from one another with variable stretches of non-transcribed spacer DNA that remains untranscribed and nucleosomal in structure, and each gene is transcribed from its promoter (black flag) to its stop signal (white flag) and not beyond. (b) depicts the situation in a lampbrush loop where the gene is transcribed from its promoter (black flag) through to *and past* its normal stop signal and into the normally non-transcribed DNA that lies downstream, thus generating very long transcription units with long transcripts that include RNA complementary to the sense strand of the gene (thick parts of the transcripts) and the nonsense DNA that lies downstream of the gene (thin parts of the transcripts).

molecular weight nuclear RNA and lampbrush loops. This 'read-through' hypothesis predicts that the number of functional genes that are expressed to form translatable RNAs may be expected to equal the number of transcription units that are active in a lampbrush set. It is a good hypothesis and it is strongly supported by evidence from a series of DR/ISH experiments on the transcription of histone genes and their associated highly repeated satellite DNA on lampbrush loops in *N. viridescens*. It says, in effect, that the only unusual feature of a lampbrush chromosome, and the very reason for the lampbrush form, is that once transcription starts it cannot stop until the polymerase meets another promoter that is already initiated or some condensed chromomeric chromatin that is physically impenetrable and untranscribable.

THREE SIMPLE EXPERIMENTS WITH ENZYMES

Is there any direct evidence in support of the read-through hypothesis for lampbrush chromosomes? Yes indeed there is, and it is of a nature that epitomizes the supreme advantages of working with these truly remarkable objects and the manner in which it is possible to do the simplest of experiments in order to obtain unequivocal information at the molecular level. This particular story goes back to 1958 when it was first shown that loops had DNA axes by dissecting lampbrush chromosomes directly into a solution of the enzyme deoxyribonuclease I (DNAase I) and then watching what happened to them. Within a few minutes, all the loops broke into thousands of little pieces. The same effect was not obtained with ribonuclease or with proteolytic enzymes. A short time later another investigator had the bright idea of timing the breakage of loops and chromosome axis by DNAase I and plotting number of breaks against time on a log scale. If, as was then rightly supposed, the chromosome axis had two chromatid strands (two double helices or four half-helices) and the loop axis had just one chromatid strand (one double helix or two half-helices) then a log plot of time against breaks for the axis should have a slope of 4 and a corresponding plot for loops should have a slope of 2 (Figure 9.11). An incredibly easy experiment, requiring only a microscope and a stopwatch, and once again, QED.

Much later, a new kind of nuclease was discovered, the restriction endonuclease, that was site specific in its cutting action on DNA. Significantly, the first availability of restriction enzymes more or less coincided with the peak time of controversy over the two classic lampbrush hypotheses. It did not take long to realize that if a loop did consist entirely of identical tandemly repeated DNA sequences, all of which possessed a particular restriction enzyme recognition site, then the loop would be destroyed by that enzyme. If, on the other hand, the DNA sequences all lacked the enzyme recognition site then the loop would be totally unaffected and would remain intact.

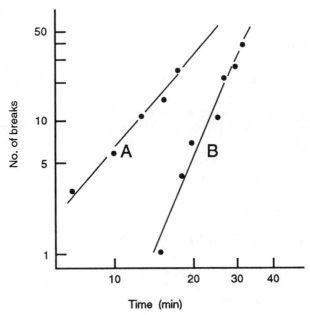

Figure 9.11 Log plots of the number of chromosome breaks as a function of time of deoxyribonuclease action. (A) Breaks in a lampbrush loop; (B) Breaks in the main chromosome axis. (Reproduced with the kind permission of Dr J. G. Gall from Gall, J.G., *Nature* **198**, 36–38, 1963.)

The experiment was set up using five enzymes and the lampbrush chromosomes from *N. viridescens*. The control enzyme was DNAase I. It destroyed all loops. Three of the other enzymes were recently isolated and only partially characterized restriction endonucleases with unknown site specificities: one of them was almost certainly *Hind*III and another *Eco*RI. The fifth enzyme was the one we refer to today as *Hae*III. Four of these enzymes destroyed all loops. *Hae*III did likewise, except that it left one set of loops completely intact. Just imagine what a remarkable observation this must have been for the investigator seated at his inverted microscope single-mindedly watching a set of lampbrush chromosomes disintegrating into smaller and smaller fragments, and then suddenly realizing that amongst this soup of destruction there was a little cluster of loops that seemed totally impervious to the enzyme's action. The loops were big ones and they mapped to the middle region of the second longest chromosome (Figures 9.12 and 9.13).

They had a curious arrangement in the sense that in some newts they regularly formed a cluster of several pairs of long loops associated with just one chromomere, whereas in other animals (of the same species) they took the form of a single immensely long loop that incorporated several tandem transcription units along its length. They came to be known as the ***Hae*III-resistant loops on chromosome II** of *N. viridescens*. Here was strong direct evidence that at least one set of loops consisted of tandemly repeated short sequence DNA. A four nucleotide sequence such as the recognition site for *Hae*III should occur by chance once in every 256 nucleotides. Here it was entirely lacking in pieces of loop axis measuring up to 100 μm, equivalent to at least 300 000 nucleotides. In a sense this was weakly encouraging news for advocates of the master/slave hypothesis, but it left the difficult matter of explaining the total susceptibility to digestion of all the other thousands of loops.

Much later again, the effects of a modern, purified and well-characterized sample of *Hae*III were tested, with appropriate controls, on the resistant loops of *viridescens* chromosome II. If the loops were present as a bunch, they detached

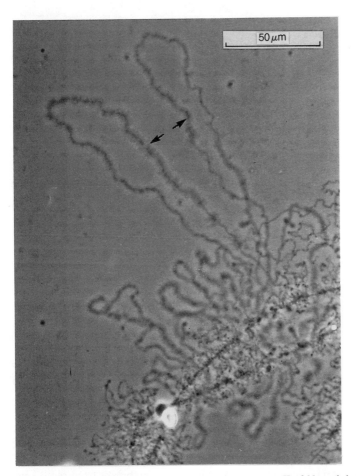

Figure 9.12 Part of the middle region of lampbrush chromosome II of *Notophthalmus viridescens* photographed very quickly after isolation into a physiologically balanced saline containing the restriction enzyme *Hae*III. One pair of the large cluster of long fluffy loops that are always found at this particular locus is indicated by arrows. (Reproduced with the kind permission of Professor H.G. Callan from Gould, D.C., Callan, H.G. and Thomas, C.A., *J. Cell Sci.* **21**, 303–313, 1976.)

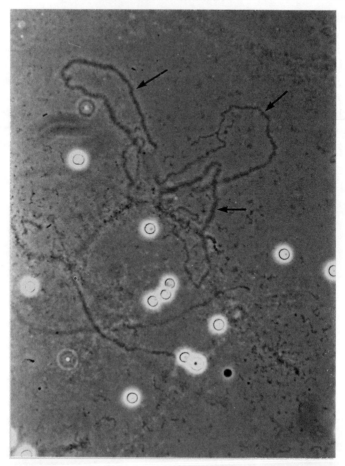

Figure 9.13 The same piece of chromosome as shown in Figure 9.11 after several minutes' exposure to the action of *Hae*III. The chromosome axes and all the 'normal' lateral loops have been reduced to countless tiny broken fragments. The large fluffy loops remain unbroken (arrows), signifying that the DNA of the axes of these loops, about 300 μm of it for each loop, lacks the recognition site for the enzyme *Hae*III (GG'CC). (Reproduced with the kind permission of Professor H.G. Callan from Gould, D.C., Callan, H.G. and Thomas, C.A., *J. Cell Sci.* **21**, 303–313, 1976.)

individually from the disintegrating chromosome and thereafter remained intact. If they were present as a single long loop consisting of several tandem transcription units (thin-to-thick segments), then breaks occurred precisely at the thin beginnings of each transcription unit, but the remainder of the loops remained intact. This is, of course, precisely as would be predicted on the basis of the currently accepted read-through hypothesis. The start of the transcription unit would be characterized by a long complex gene sequence that would almost inevitably include the *Hae*III recognition site. The remainder of the loop would consist entirely of repeat sequences that lacked the *Hae*III site.

SOME MORE QUESTIONS ABOUT LAMPBRUSHES

Are there any DNA sequences on lampbrush chromosomes that are never transcribed? Probably not. There are certainly some large blocks of tandemly repeated short DNA sequences that condense into large lumps of hetero-chromatic material with no associated loops. However, careful investigation usually reveals some parts where even the most nonsense-like sequences are transcribed, presumably by read-through from interspersed functional gene sequences.

Only a small fraction of the entire DNA of a loop–chromomere complex forms the transcription unit that makes the loop. What about the rest of the DNA? How many potential promoter sites are there in a chromomere? Is the part that makes the loop preferentially or randomly selected? Can we liken the process to specifically picking out a particular stretch of chromomeric DNA for transcription, the same piece at the corresponding locus in every egg of every individual of a particular species. Or should we liken it to inserting a crochet hook into a ball of wool and pulling it out with a random loop on the end of it? We do not know the answers to any of these questions, but it would not be hard to design experiments in search of them.

Why do loops have different morphologies that are heritable, locus specific and sometimes species specific? Their basic organization is, for the most part, the same, consisting of different levels of folding and secondary structure imposed on the initial nascent RNA transcript. The loop matrix is a site of processing, cleaving and packaging of nuclear RNA, so most of the variation in gross structure may be expected to reflect different modes of binding and interaction involving quite a wide range of proteins and RNAs. Callan and Lloyd's classic 1960 paper in the *Philosophical Transactions of The Royal Society* should be examined to see just how variable loops can be, whilst at the same time being faithfully recognizable from one preparation to another within an entire population of animals.

Do lampbrush chromosomes look the same in all animals? Naturally not. The lengths of lampbrush chromosomes at the time of their maximum development is in quite strict agreement with the relative lengths of the corresponding mitotic metaphase chromosomes from the same species. The overall lengths of lampbrush chromosomes are broadly related to genome size and chromosome number. Newts, with genomes of between 20 and 40 pg have much longer lampbrushes than frogs with genomes of 3–10 pg (Figure 9.14).

Some lampbrush chromosomes have long loops and others have very short ones. We have seen that the transcription units of lampbrush chromosomes are unusually long because they include interspersed repetitive elements of the genome. Structural genes in large genomes are more widely spaced, interspersed

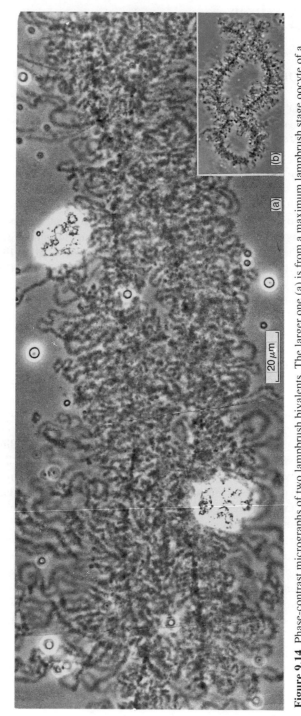

Figure 9.14 Phase-contrast micrographs of two lampbrush bivalents. The larger one (a) is from a maximum lampbrush stage oocyte of a Central American salamander, *Bolitoglossa subpalmata*, that has a genome size of 87 pg of DNA (haploid chromosome number 13), probably the biggest lampbrush chromosome in the world! The smaller lampbrush bivalent (b) is from a reptile, *Bipes canaliculatus*, that has a genome size of 2 pg of NA (haploid chromosome number of 11 macrochromosomes plus 11 microchromosomes). Only half the *Bolitoglossa* bivalent is shown. These two bivalents are shown at precisely the same magnification, and together they provide a clear demonstration of the way in which lampbrush chromosome size and loop length varies according to genome size. (Reproduced from Macgregor, H.C., *Heredity* **44**, 3–35, 1980.)

with non-coding DNA, than in small genomes. One might therefore expect lampbrush chromosomes from large genomes to have longer loops (transcription units) than those of smaller genomes, and this is precisely what has been observed. That is not, however, the whole story. It has been known for some time that some very large genomes, like that of the North American mud puppy (*Necturus maculosus*) with a genome that is at least four times as large as the crested newt and nearly 30 times as large as that of *Xenopus laevis*, have lampbrush chromosomes with very short loops. It is also known that, whereas *Xenopus* lampbrush chromosomes have small, stumpy and generally uninspiring loops in normal circumstances, if we do something to inhibit post-transcriptional processing of the loop RNA, then the loops become much longer and the general appearance of the entire lampbrush chromosome set changes dramatically.

One final point in relation to loop morphology serves, yet again, to illustrate how we can extract molecular information from lampbrushes simply by looking at them with a microscope. Many of the very long loops that we see in lampbrushes from animals with large genomes show multiple, tandemly arranged thin–thick segments (transcription units). The remarkable thing about these situations is that the individual transcription units within one loop can have the same or opposite polarities and can be of the same or different lengths (Figure 9.15). This observation, perhaps better than any other, tells us that it is really the transcription unit that is the ultimate genetic unit in a lampbrush chromosome and not the loop–chromomere complex as was once thought.

Why do lampbrush chromosomes exist at all? They are characteristic of eggs that develop rather quickly into complex multicellular organisms independently

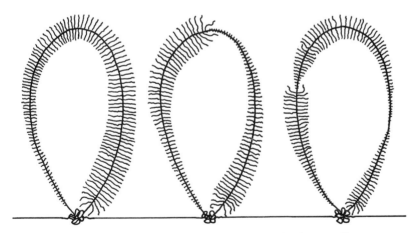

Figure 9.15 The various arrangements of transcriptions units that actually occur on lampbrush chromosomes. On the left is a loop consisting of a single transcription unit. In the middle is a loop consisting of two transcription units of the same size and with the same polarity. On the right is a loop with four transcription units of different sizes and different directions of polarity.

of the parent. A frog's egg, for example, is fertilized, deposited by the mother and then develops into a complex tadpole within a few days. Much of the information for this process in the form of polyadenylated messenger RNA, most of the ribosomes for protein synthesis and all of the nutrient raw materials are laid down during oogenesis through activity of lampbrush chromosomes and amplified ribosomal genes and the accumulation of yolk proteins imported from the liver. Lampbrushes may therefore be regarded mainly as an adaptive feature that has evolved to preprogramme the egg for rapid early development. The fact that they are not developed in mammalian eggs could be regarded as a primitive feature that is consistent with the relatively slow pace of mammalian development. A frog's egg, for example, will have completed gastrulation and be well advanced in the differentiation of its central nervous system by the time a human egg has reached the eight cell stage.

RECOMMENDED FURTHER READING

Callan, H.G. (1986) *Lampbrush Chromosomes*. Springer-Verlag, Berlin.

Callan, H.G. and Macgregor, H.C. (1958) The action of deoxyribonuclease on lampbrush chromosomes. *Nature* **181**, 1479–1480.

Diaz, M.O. and Gall, J.G. (1985) Giant readthrough transcription units at the histone loci on lampbrush chromosomes of the newt *Notophthalmus. Chromosoma (Berl.)* **92**, 243–253.

Gall, J.G. (1963) Kinetics of deoxyribonuclease action on chromosomes. *Nature* **198**, 36–38.

Gall, J.G., Diaz, M.O., Stephenson, E.C. and Mahon, K.A. (1983) The transcription unit of lampbrush chromosomes. In *Gene Structure and Regulation in Development*. Alan R. Liss, New York, pp. 137–146.

Gould, D.C., Callan, H.G. and Thomas, C.A. (1976) The actions of restriction endonucleases on lampbrush chromosomes. *J. Cell Sci.* **21**, 303–331.

Macgregor, H.C. (1980) Recent developments in the study of lampbrush chromosomes. *Heredity* **44**, 3–35.

Macgregor, H.C. (1987) Lampbrush chromosomes. *J. Cell Sci.* **88**, 7–9.

Solovei, I. Gaginskaya, H., Allen, T. and Macgregor, H.C. (1992) A novel structure associated with a lampbrush chromosome in the chicken *Gallus domesticus. J. Cell Sci.* **101**, 759–772.

DNA sequences and genome evolution | 10

In nearly all of the previous chapters there has been some mention of the DNA sequences that make up the chromosomes of higher organisms. We have encountered the sequences that code for things like ribosomal RNA, the short repeated sequences that make up the bulk of the DNA in heterochromatin, specific sequences that seal the ends of chromosomes, others that are involved in the events of synapsis and crossing over and sequences that mark the beginnings and ends of genes that are to be read and transcribed by RNA polymerase.

THE CLASSIFICATION OF DNA SEQUENCES

For the sake of convenience and as a part of the language of molecular biology, chromosomal DNA sequences have been categorized with regard to their structural and functional properties. The categories are not entirely real, and the overlaps and grey areas between them are considerable, but they are nonetheless useful. They are as follows:

- *Single copy sequences*, complex sequences represented just once per haploid genome, many of them genes that code for functional polypeptides.
- *Moderately repeated sequences,* relatively simple sequences of variable length that may be represented by up to a few thousand repeats, generally non-coding and quite widely scattered throughout the chromosome set.
- *Mobile elements,* a subset of moderately repeated sequences that, as the name suggests, have the ability to move from place to place among the chromosomes. Some of them are transcribed.
- *Functional repeated sequences* (sometimes called multigene families), complex sequences that code for specific functions and are represented by several hundred to a few thousand repeats, e.g. ribosomal and histone genes.
- *Highly repeated sequences* (HR DNA), short and highly variable non-coding

sequences, usually represented by several thousand to several million copies per haploid genome, mostly clustered into large heterochromatic blocks that are identifiable as Giemsa C-bands.

- *Centromere sequences* (CEN), specific DNA sequences that have been identified at or around the centromere regions of yeast chromosomes and presumed also to be present in different forms at the centromeres of higher eukaryotes.
- *Telomere sequences,* short simple sequences that are always present at the extreme ends of eukaryotic chromosomes.

There are many other smaller categories and subcategories, but we need not concern ourselves with them for the moment. Instead let us concentrate on just two major classes of DNA sequence, the functional repeated and the highly repeated DNAs, and see what we can learn by adopting the comparative approach and examining the variation in these types of DNA from one species to another.

VARIATION IN REPEAT SEQUENCE DNA

Are there differences in the amounts and sitings of HR DNA, and, if so, are they reflected in phenotype?

In *Drosophila*, anything goes. The amount of HR DNA can be almost doubled, and blocks of heterochromatin that are rich in HR DNA sequences can be moved around from place to place in the karyotype without apparent effect on the development, the metabolism or the appearance of the fly. From one species of *Drosophila* to another, HR DNA can make up from 0 to 60% of the genome. It can be concentrated into a few heterochromatin blocks on one or two chromosomes or it can be scattered in smaller blocks amongst all the chromosomes. The sizes of the monomer sequences that are highly repeated can vary from 5 to 360 base pairs (bp). The sequences themselves make absolutely no sense.

The situation elsewhere in the animal kingdom is the same. Amongst the kangaroo rats (genus *Dipodomys*) species differ by having between 5% and 60% of their genomes occupied by HR DNA, and there are no obvious associated phenotypic differences. The mouse has 10% of its genome as a 234-bp repeat. The rat has 3% of its genome as a 370-bp repeat. Choose whatever group of species you like and you will find HR DNAs of all sizes and arrangements and representing a wide range of percentages of the genome. There are even examples of closely related species that have widely different amounts of heterochromatin, making corresponding differences in the sizes and shapes of chromosomes: and yet these species are capable of extensive hybridization in the wild, the differences in heterochromatin and its constituent HR DNA being no barrier whatever to the production of viable F1 hybrids.

What about differences in the amounts and sitings of functional repeated sequences? The 5S, 18S and 28S ribosomal genes and the histone genes provide useful examples within this category.

In lower eukaryotes, the yeasts and the slime moulds, all the ribosomal RNA genes are transcribed together. Yeast has about 100 copies of this complex located on just one of its chromosomes.

In higher eukaryotes, the 5S genes are quite separate from the other two ribosomal genes. The number of repeats of the 5S gene per haploid genome varies from 100 in yeast to around 300 000 in the North American newt *Notophthalmus viridescens*. *Xenopus laevis* has 24 000 repeats; *X. borealis* has 600 repeats.

In humans and *Drosophila* all the 5S gene repeats are clustered together at just one site on one chromosome. In other animals they may be scattered amongst several sites. *Xenopus laevis* has 15 different chromosomal loci, all of them near the ends of chromosomes, at which there are clusters of 5S sequences.

The same general picture is found for the 18S+28S ribosomal genes. *Xenopus laevis* has one locus, *Notophthalmus* has three, humans have five, gibbons have one. Number of repeats varies widely and, as we have already seen in Chapter 8, it even differs between individuals within one population of a single species.

The genes for histones, the five proteins that complex with DNA to make chromatin, are usually clustered together into one large repeating unit, but the order in which they are arranged in this basic cluster is different from one species to the next. *Drosophila melanogaster* has about 100 repeats, all of which are located in one short region of the so-called left arm of chromosome 2. Humans have about 40 repeats with the genes for histones H3 and H4 on chromosome 1, others on chromosome 7 and still others elsewhere. The same general pattern is found in mice and chickens, which suggests that it was established before the evolutionary divergence of birds and mammals. *Notophthalmus* has just a single histone site at which there are 600–800 repeats, each about 9 kilobases (kb) long and separated from the neighbouring repeat by variable lengths of non-coding spacer DNA belonging to the HR DNA category.

All this must surely serve to fortify an impression of the eukaryotic genome as a largely haphazard assembly of DNA sequences generated by random events that are kept in check only by meiosis and natural selection. Later we shall take a brief look at the kinds of things that can happen to a piece of a genome when it is released from meiotic constraints and exempted from major selective forces.

At the level of the chromosome all kinds of variations and changes must somehow be accounted for. Whole chromosomes become shorter or longer. Giemsa C-bands come and go. Multigene families expand or contract. Genes move from one location to another. How does it all happen, and what regulates it, for regulation there must be? Otherwise how shall we account for constancy of chromosome number and form within species, for the basic similarity of karyotypes within whole classes of animals and, more particularly, for the precise maintenance of chromosome number, relative size and shape in some genera despite big changes in chromosome size? Within the individual chromosome things may seem to be random and haphazard. Within the karyotype the rules are clearly much stricter.

AGENTS OF CHANGE

We have already met some of the blunter instruments of genome change. Inversions and translocations are obviously important. They originate in single individuals and, if they are selectively neutral or confer some advantage, they spread throughout the population and become fixed. It is important to think carefully about how this might happen, with special regard to the problems of meiosis and crossing over in structural heterozygotes, as described in Chapter 3.

Amplifications, deletions and duplications are also important at every level from individual DNA sequences up to whole segments of chromosomes. One of the most astounding examples of amplification was discovered in the course of a programme of research into the actions of drugs used to control the growth of tumours. Some cells in tissue culture are able to survive stepwise increases in the concentration of the antifolate drug **methotrexate** (used in some regimens of cancer chemotherapy) because they dramatically overproduce the target enzyme **dihydrofolate reductase (DHFR)**. The overproduction results from greatly elevated levels of production of the messenger RNA for DHFR. This follows a massive amplification of the genes for DHFR, which is accompanied by the appearance of new chromosome forms. Either there is a measurable increase in the length of the chromosome arm that carries the DHFR gene and/or there is a sudden appearance of lots of very small new chromosomes that lack centromeres and consist largely of amplified DHFR gene sequences. Truly an extraordinary happening, but, we must ask, could such a thing happen in a heritable and selectively advantageous manner? Could the molecular mechanisms underlying methotrexate resistance be ready and available at all times and in all cells as potential agents of genome change?

Deletions are commonplace, and the ones that contribute significantly to genome change or karyotype evolution usually involve non-coding sequences or multigene families, for obvious reasons: deletion of structural genes or functional coding sequences is unlikely to be selectively neutral or advantageous.

Duplications are another matter altogether. The earliest description of a duplication at the level of the chromosome, published long before we knew anything about molecular mechanisms (Bridges 1936), provided clear and simple evidence of one of the most important agents of genome evolution. It was the cytological characterization of the BAR mutation in *Drosophila melanogaster* in terms of the banding patterns on polytene chromosomes.

The BAR mutation affects the shape of the eye in the adult fly. Normal eyes are round; BAR eyes are, as the name suggests, bar-like. The BAR condition is unstable and occasionally it alters to a condition known as ULTRABAR, where

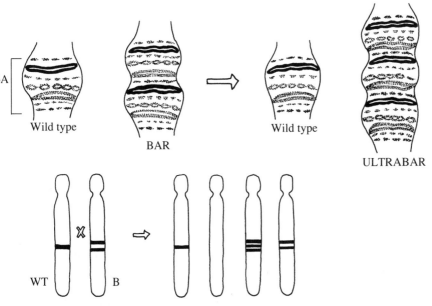

Figure 10.1 The 16A region of the *Drosophila melanogaster* X chromosome in a wild-type, BAR and ULTRABAR fly, and a diagram to show how unequal crossing over in a female heterozygous for BAR could generate an X chromosome with three tandem copies of the BAR locus and an ULTRABAR phenotype in the following generation.

the eye is reduced to a thin vertical strip. BAR arose as a straight duplication of a cluster of about seven bands on the X chromosome (Figure 10.1).

Individuals heterozygous for BAR have one BAR locus on one chromosome and two in tandem on the other. It is with the production of ULTRABAR that a major secret of genome evolution begins to unfold. ULTRABAR arises as a consequence of **unequal crossing over** between a chromosome with one BAR locus and one that has two BAR loci in tandem. The outcome is a chromosome that lacks the BAR locus altogether and one that has three BAR loci in tandem (Figure 10.1). Could *this* be a mechanism that is ready and available at all times and in all cells as a potential agent of genome change and karyotype evolution, for, if it were, it would be a very powerful agent indeed.

The logic is as follows. The genomes of higher organisms are characterized by repeat sequence DNA. Differences in genome sizes between taxonomic groups, species, genera, orders, classes, etc., are not in respect of coding sequences or structural genes but are largely attributable to different amounts, diversities and distributions of non-coding repeated DNA sequences. The differences are substantial, they are heritable and they can arise quite quickly. If we are to account for them mainly on the basis of unequal crossing over, primed by gene duplication and augmented by amplification, deletion and structural rearrangement, then we have to satisfy ourselves on two matters.

First, we have to be convinced that exchanges, cross-overs, can take place between homologous chromosomes during mitosis. Exchanges that happened only in meiotic prophase would be too infrequent to account for the observed levels of variation and rates of change. For exchanges to have a significant impact on the genome and for them to be heritable, they must take place in either the oogonial or the spermatogonial mitoses, producing whole clones of gametes that are capable of passing the change on to the next generation. Second, we have to have some evidence that exchanges of this kind can be unequal in the same sense as the BAR to ULTRABAR event involved an unequal cross-over.

SISTER CHROMATID EXCHANGE

The occurrence of sister chromatid exchanges (SCEs) was first noted by J.H. Taylor in his early studies of chromosome duplication employing tritiated thymidine and autoradiography (see Chapter 7 and Figure 7.1). Much later, at a time when rapid developments were taking place in the differential staining of chromosomes with fluorochromes and Giemsa (see Chapter 2), it was dis-covered that if cells were grown for two cell cycles in tissue culture in the presence of 5-bromodeoxyuridine (BrdU, a base analogue that replaces thymidine as the DNA replicates at S-phase) then the two sister chromatids in the second metaphase were differentially stained. If the stain was a fluoro-chrome, then one chromatid was bright and one was faint. If the stain was Giemsa, one chromatid was dark and one was lighter in shade (Figure 10.2).

The difference derives from the fact that after two rounds of replication in BrdU, one of the sister chromatids has BrdU in just one of its constituent polynucleotide strands (unifilar substitution), whereas the other has BrdU in both strands (bifilar substitution). BrdU discourages binding of either fluorochrome or Giemsa, so that the bifilarly substituted chromatid will be the fainter or the lighter of the two. If this proves hard to understand, then make a drawing like that shown in Figure 7.1, bearing in mind that we are substituting BrdU for tritiated thymidine and the BrdU is available to the cells at both the first *and* the second S-phase. BrdU, then, allows us to stain sister chromatids differentially. If there has been an exchange between these chromatids sub-sequent to the previous S-phase then it will show clearly as a transfer of differentially stained segments between chromatids (Figure 10.2). It must be added that BrdU itself tends to induce chromatid breaks that result in SCEs. However, a natural level of SCE can be deduced by growing cells in a range of BrdU concentrations, plotting frequency of SCE per mitosis against BrdU concentration, and then extrapolating back to the frequency of SCEs that would be expected at zero BrdU (Figure 10.3).

Procedures of this kind demonstrate beyond doubt that sister chromatid exchange is a regular feature of mitosis in plant and animal cells. The matter of *unequal* SCE can only be approached indirectly through precedent and

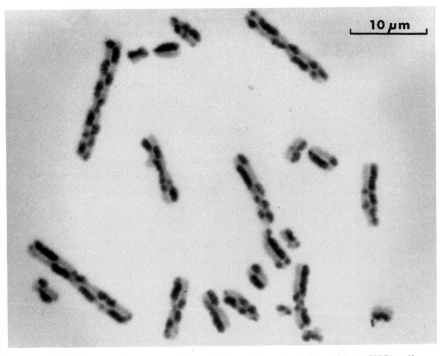

Figure 10.2 Mitotic chromosomes from a Chinese hamster tissue culture (CHO) cell showing multiple sister chromatid exchanges. The cells were grown in the presence of BrdU and the chromosomes were stained with the dyes Hoechst and Giemsa in such a way as to reveal the differentially light staining of chromatids in which thymidine has been replaced by BrdU. In this particular instance multiple sister chromatid exchanges had been induced by treatment of the cells with the drug cyclophosphamide. (Photograph kindly provided by Dr Sheldon Wolff and Judy Boycote of the Laboratory of Radiobiology and the Department of Anatomy, University of California at San Francisco.)

probability. The precedents are that it happens with BAR and it has been shown to happen within the cluster of ribosomal genes in yeast. In both these cases, unequal exchanges between homologous chromosomes, and in the case of yeast also between sister strands, have been unequivocally demonstrated. The yeast experiment, carried out by Petes and his associates in 1980, is a particularly beautiful one, involving some very crafty genetic engineering to make chromosomes with different arrangements of ribosomal genes, but it is too complex to describe in detail here.

The probability that SCEs are unequal can be assessed by producing a model and testing it to see how effective it is in generating the kinds of real-life situations that we find in modern genomes. In other words, if unequal SCE is to be afforded the status of a major instrument of genome evolution then its credentials had better match up to requirements.

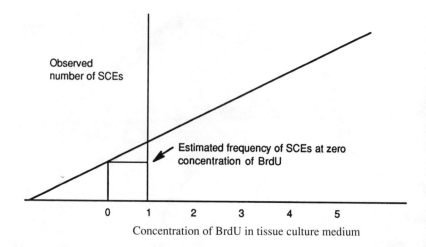

Figure 10.3 The increase in frequency of sister chromatid exchanges with concentration of BrdU, and extrapolation back to zero concentration of BrdU in order to estimate the frequency of sister chromatid exchange in the absence of the compound.

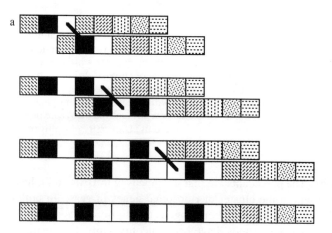

Figure 10.4 A simple model of the generation of repeat patterns in a replicating linear array of units by replication, relative longitudinal displacement of replicas and unequal crossing over between replicas. In (a) we see a line of eight patterned squares that may be considered to represent a chromosome made up of eight genetic units. This replicates, the replicas displace and a cross-over (black diagonal bar) occurs between the third unit in one copy and the second in the other. Two new chromosomes are generated: one has six units while the other has ten. A repeat of the black/white pair of units has been generated in the longer of the two chromosomes. The same principles are applied through two more rounds of replication and unequal crossing over (in both cases artifically selecting for the longer of the two products of unequal cross-over), and a whole range of repeat patterns involving the black/white units are generated.

Constructing a model is easy. A simple one is illustrated in Figure 10.4. It begins with a linear array of eight units representing a chromosome (Figure 10.4a). The chromosome then replicates to produce two identical chromatids.

The chromatids are then displaced longitudinally with respect to one another, after which a cross-over occurs involving break-points that are not homologous (Figure 10.4). This generates two new chromatids, one of which is longer than the other.

The longer one has gained a duplication: the shorter one has suffered a deletion. The model continues with replication of the longer chromatid, lengthwise displacement of the two daughter chromatids and the occurrence of a nonhomologous cross-over. Once again, the longer of the two chromatids has gained a duplication; the shorter has suffered a deletion. If we now take this model through a third round of replication, displacement and unequal crossing over, we end up with a chromatid that has three different levels of repeat, it has repeats within repeats and it has palindromic repeats. All of these are features that are commonly found amongst repeat sequence families in the genomes of eukaryotes.

The model shown in Figure 10.4 is, of course, rather too simple, but it nevertheless conveys in principle how unequal SCE could generate the kinds of repeat that we see in real life and could cause chromosomes and genomes to expand and contract. The beauty of this model is that it is readily amenable to testing by computer. The starting point, a random linear array of units, some of which may be tandemly duplicated, is straightforward. The rules are easy to define. The repeating cycle of events could not be simpler. We can even put in some real parameters, since we know about genome and chromosome sizes, we know about cell cycle times and we know something about the frequency of sister chromatid exchange. The results of such a test are nothing short of spectacular (Smith 1976). They show that, within the range of normality, unequal sister chromatid exchange is a probable and potent source of most of the arrays of repeated DNA sequences that have been identified by DNA sequencing techniques. The conclusion that it is an instrument of genome expansion is almost inescapable. If it happens at all then it will, as we have seen in our simple model in Figure 10.4, generate large and small products. On average it seems likely that there will be positive selection for the larger products, since selection of the smaller products would carry a risk of loss of important functional gene sequences. Accordingly, there should be a general tendency for chromosomes to grow.

WHAT KEEPS GENOMES IN CHECK?

What are the constraints on genome change? Clearly there must be some. Genomes do not just keep on growing larger. Karyotypes are conservative, suggesting that they are balanced and integrated systems that have evolved under the influence of natural selection. What keeps them in check?

Three major factors are of prime importance. Meiosis is the most important of all. It requires homology, so that in diploid bisexual organisms what happens on one chromosome must not affect its homology with its partner to an extent that will interfere with meiotic pairing, crossing over and disjunction. The 'meiotic sieve' is extremely unforgiving! Lots of organisms can make F1 hybrids if they are physically able to mate and their genomes, when combined, are capable of supporting growth and development. Very few can make F2s, for that means putting the combined hybrid set of chromosomes to the meiotic test so as to make functional gametes.

Linkage is perhaps the next most important factor. We have seen that to some extent the position of a gene or multigene family in the karyotype is un-important, at least with regard to growth, development and metabolism of the individual. It would be misguided to suppose, however, that we can scramble a karyotype and get away with it at the population or species level. Consider two isolated and independent populations from one species that have diverged karyotypically mainly with respect to the distributions of their chiasmata in meiosis. One has evolved to have all of its chiasmata near the ends of its chro-mosomes, the other has them in the chromosome arms away from the ends. They hybridize and the F1 hybrids are fine. They grow into normal adults and they are fertile. The two genomes would seem to be entirely compatible, until the F2 or F2-backcross generation comes along. In these, we are seeing animals that are the result of fusions between gametes that were the product of meiosis in the F1 hybrids. Where did the chiasmata form in these meiotic divisions? Whatever happened, one or other of the original parental genomes was subjected to unfamiliar patterns of crossing over, and that inevitably means break-up of favourable gene linkages that evolved in the original parent type. Scarcely surprising, then, that the F2s suffer arrest during early development and fewer than half the F2 back-crosses develop at all. This experiment, carried out by a team of Australian investigators with some extremely interesting isolated populations of Australian grasshoppers, supported the concept of the **coadapted genome** (Coates and Shaw 1982, White 1978). The results are considered to signify that the placement of chiasmata in regions of normally low re-combination disrupts the internal coadapted genetic environment in both chromosomal forms, resulting in non-viable recombinant progeny in the next generation.

The third factor that keeps genomes in check is more hypothetical. We have seen that genomes come in different sizes. It is not hard to imagine that there will be a positive correlation between genome size and cell volume, if only on the basis that bigger cells are needed to make room for bigger chromosomes. There is indeed such a correlation. It is good and it is linear (Figure 10.5).

What is more, although it is much harder to demonstrate convincingly, there is a similar correlation between genome size and the time that it takes a cell to

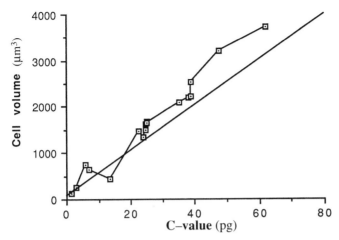

Figure 10.5 The simple linear relationship between genome size (C-value) and cell volume for red blood cells from 16 species of amphibian. (Reproduced from Horner, H.A. and Macgregor, H.C., *J. Cell Sci.* **63**, 135–146, 1983.)

pass through one complete mitotic cycle. The difficulties with comparative studies of cell cycles is that within one organism different types of cell have quite widely different cell cycle times, so when comparing species it is essential to be sure that one is measuring the cycle times of similar cells under as closely identical conditions as possible. The hypothesis is that genomes have a natural tendency to expand by accumulation of repeated sequence DNA. Genome expansion will be kept in check by selective forces that favour more rapid growth and development, such as is possible with smaller, more rapidly dividing cells.

Two final questions seem appropriate to round off this chapter. First, amongst all the millions of nucleotides and sequence arrays in a eukaryotic genome and amidst all the changes that are constantly being tried, tested, adopted or discarded in the evolutionary process, has anything escaped change? The answer to this is yes, some things have been conserved and some preserved. **Conservation** implies that something will remain unchanged if it has proved to be efficient and essential and has not ever been placed in competition with a variant form that represents something better and will confer a competitive advantage on the organism. The ribosomal RNAs, tubulin, myosin, haemoglobin and histone are all examples of things that are specified by genes that have remained largely unchanged for millions of years and are found in wide ranges of living organisms belonging to large taxonomic groups. **Preservation** implies that something that is neutral with respect to natural selection has escaped change through being tucked away in a corner of the genome where it is undisturbed by rearrangements, duplications, cross-overs, deletions and all the rest of the scrambling agents that are continually at work.

Undoubtedly there are some parts of our own genomes and karyotypes that are very old indeed. We have already learned of several examples, such as linkage of genes on chromosome 1 in primates that diverged about 5 million years ago (Chapter 2). There are shared linkage groups on chromosome 21 in humans and chromosome 16 in mouse (Chapter 11); primates and rodents are some 85 million years apart. It is clear that autosomal segments are conserved or preserved between eutherian and metatherian mammals that diverged 130–150 million years ago, and even between eutherians and prototherians, which diverged 150–170 million years ago (Human Gene Mapping 11, 1991). So far, almost nothing is known about possible conservation or preservation of linkage groups between different classes of vertebrates. Indeed, the striking differences between, for example, the karyotypes of mammals and birds must surely lead us to suppose that vast rearrangements affecting the order of gene loci and DNA sequences must have happened in the course of time and very little of any ancient genome can possibly remain. Nevertheless, it has been reported that the albumin and vitamin D-binding protein genes have remained linked in chickens, humans and horses (Juneja *et al.* 1982). Therefore, these species share at least one chromosome segment that has been conserved (or preserved) since the time of divergence from a common ancestor, approximately 300 million years ago.

THE STORY OF A LIBERATED CHROMOSOME

Lastly, what happens to a chromosome in the karyotype of a diploid bisexual organism if it is liberated from the constraints imposed by the meiotic process? One such example has been characterized. About 20 million years ago in a small pond somewhere on what is now the continent of Europe there were some newts, probably two or three hundred of them, amongst which was the animal that was to be the ancestor of all the marbled (*Triturus marmoratus*) and crested newts (*T. cristatus*) of modern times. This animal had 12 pairs of metacentric chromosomes. In one of the cells in its testis a major event took place on both homologues of the long arm of one of the larger chromosomes. The event was an unequal reciprocal translocation, such that a short piece of one chromosome exchanged with a long piece of the other (Figure 3.25b).

This made two chromosomes that were markedly different in length. That cell then continued to divide and a clone of cells (spermatogonia) was produced, all of which carried the new translocated chromosomes. The cells then entered meiosis, and in the region of the reciprocal translocation the chromosomes failed to synapse and no chiasmata formed. The short arms of the chromosomes synapsed normally and chiasmata did form, so disjunction of the whole chromosome was unaffected. Gametes were made, they fertilized eggs and a generation of newts emerged that carried one normal chromosome and one translocated one. In the next generation, the first individuals that had two translocated chromosomes were produced. Those that had two of the same failed to develop properly and died. Those that had one of each developed normally. For reasons that remain a mystery, all the newts that had two of the original untranslocated chromosomes disappeared and the population gradually came to include only animals that had one each of the two translocated forms. Subsequently, newts from this population migrated to other ponds, mixed and mated with other individuals and, in the course of several million years, spread the rearranged chromosomes into a total of five highly successful species.

These translocated chromosomes were something quite extraordinary. They were large submetacentric chromosomes that consisted of a normal short arm that was the same in both members of the pair. This arm, together with a short region just across the centromere, was capable of proper meiotic synapsis and chiasma formation, which ensured that the chromosomes behaved themselves absolutely normally at each meiotic division. Attached to the short arms were long arms that never paired and were never involved in meiotic recombination. The genes and DNA sequences in these long arms were therefore entirely free to rearrange, duplicate, transpose, mutate, do anything, just so long as it did not interfere with proper expression of the functional genes in the region. They did all of these things. The arms grew longer through the accumulation of multiple repeats of non-coding sequences. This made the chromosomes the longest in the karyotype and also made them more conspicuously submetacentric than any other chromosome in the entire genus, so that when we discovered them, 20 million years after they first evolved, they were given number 1 status in the species idiogram, the biggest chromosome pair of the 12. The asynaptic regions of the long arms are now strongly heterochromatic and, unlike the remainder of the chromosome set, they stain darkly with Giemsa to form a coarse pattern of Giemsa C-bands.

When we look closely at these asynaptic regions, four features are particularly evident. First, there is a very large amount of highly repetitive DNA. Second, the arms seem to have become a sink for a wide range of DNA sequences, including small clusters of some multigene families in addition to the main clusters of these same families elsewhere in the karyotype. Indeed, almost every DNA sequence that has ever been *in situ* hybridized to the chromosomes of crested or marbled newts has been found to bind to the long arms of chromosome 1 as well as to somewhere else. Third, the long arms of the two, originally homologous, chromosomes 1 are now completely different. They are of different sizes, they have different Giemsa banding patterns, they have different loops when examined in the lampbrush condition and they have different gene arrangements (Figure 10.6).

They have only one thing in common: they are both loaded with highly repetitive DNA. Lastly, these two chromosomes are highly variable from species to species, even from population to population within a species, and perhaps even from one individual to another. The Giemsa banding pattern varies, the lampbrush loops vary, the locations of small vagrant clusters of ribosomal genes vary. Variation of this kind spells only one thing: the chromosomes must be

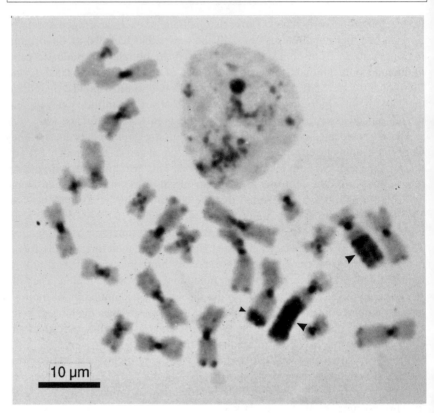

Figure 10.6 A Giemsa C-banded metaphase chromosome complement from a male marbled newt (*Triturus marmoratus*). The two large arrows indicate the two strikingly heteromorphic long arms of the longest chromosome pair in the set, no. 1. These arms are strongly Giemsa C-positive, signifying their high content of heterochromatin based on repeated DNA sequences. They are of different lengths and they have distinctly different patterns of staining with Giemsa. The large size of these arms makes the chromosomes no. 1 more submetacentric than any other chromosome in the set. The small arrow indicates the Y chromosome terminal Giemsa band. (Reproduced with the kind permission of Dr S. Sims and John Wiley & Sons from Macgregor, H.C. and Varley, J.M., *Working with Animal Chromosomes*, 2nd edn, 1988.)

changing at a rate that is vastly in excess of that which is tolerable in other parts of the karyotype, parts where changes must always be precisely duplicated, homologue for homologue, or meiosis will fail and the changes will be filtered into oblivion.

RECOMMENDED FURTHER READING

Clarke, B.C., Robertson, A. and Jeffreys, A.J. (1986) *The Evolution of DNA Sequences*. Proceedings of the Royal Society of London Discussion Meeting March 1985. The Royal Society, London.

Coates, D.J. and Shaw, D.D. (1982) The chromosomal component of reproductive isolation in the grasshopper *Caledia captiva*. *Chromosoma (Berl.)* **86**, 509–531.

Hoelzel, A.R. and Dover, G.A. (1991) *Molecular Genetic Ecology*. IRL Press, Oxford.

Human Gene Mapping 11 (1991) *Cytogenet. Cell Genet.* **58**, 1–2200.

Juneja, R.K., Sandberg, K., Anderson, L. and Gahne, B. (1982) Close linkage between albumin and vitamin D binding protein (Gc) loci in chicken: a 300 million year old linkage group. *Genet. Res. Camb.* **40**, 95–98.

Macgregor, H.C. (1991) Chromosome heteromorphism in newts (*Triturus*) and its significance in relation to evolution and development. In *Amphibian Cytogenetics and Evolution*. D.M. Green and S.K. Sessions (Eds.). Academic Press, San Diego. pp. 175–196.

Smith, G.P. (1976) The evolution of DNA sequences by unequal crossover. *Science* **191**, 528–535.

Strachan, T. (1992) *The Human Genome*. Bios Scientific Publishers, Oxford.

White, M.J.D. (1978) *Modes of Speciation*. W.H. Freeman & Co., San Francisco.

Human clinical cytogenetics | 11

The history of the development of human cytogenetics was outlined in Chapter 2. It is, of course, an extremely rapidly expanding and exciting field of biomedical science that is represented, on the one hand, by frontiers of science programmes in molecular biology, in areas such as human genome mapping, gene therapy and the molecular basis of inherited disorders of metabolism and development, and, on the other hand, by clinical programmes mainly directed at the manage-ment of the more common disorders that are known to be chromosomally based. This chapter is about the latter of these two areas of work. Its main purpose is to offer a brief, behind-the-scenes look at what goes on in the very practical and down-to-earth world of a clinical cytogenetics service laboratory.

Clinical cytogenetics is a field of general importance in the sense that we are naturally interested in our own chromosomes. Chromosome defects that are associated with socially and medically important problems have driven us forward in the development of techniques that we might not otherwise have bothered with if they were only applicable to non-human matters. It is a field of professional interest for undergraduates in the sense that employment oppor-tunities for trainee cytogeneticists are generally quite good. What is more, a training in practical clinical cytogenetics will generate a range of transferable skills that are likely to be valuable in several other fields of biomedical science.

The basic core technology of clinical cytogenetics is G-banding of chro-mosomes from cells grown in tissue culture. By far the greater part of the workload in a cytogenetics laboratory is tissue culture, G-banding and chro-mosome interpretation with the resolution available from metaphase prep-arations that show about 500 bands per haploid set.

CONVENTIONS FOR DESCRIBING HUMAN CHROMOSOMES

There is, as one might expect, a tight internationally agreed system for human cytogenetic nomenclature (ISCN 1985) that has been updated three times since 1971. It is based primarily on G-band patterns, although it also includes information on quinacrine (Q-) banding, reverse (R-) banding and C-banding. In describing a human karyotype the following rules apply.

1. First, the **entire chromosome number** of the individual is given. For example:

    46

2. The **sex chromosome constitution** is then stated with a comma after the chromosome number. For example:

    46,XY

3. If there is an abnormality, then the **number of the abnormal chromosome** is given next, once again, after a comma. For example:

    46,XY,16

4. Then comes the **arm of the abnormal chromosome** that carries the abnormality, '**p**' being the short arm and '**q**' the long one. For example:

    46,XY,16q

5. Lastly, we specify the nature and the location of the abnormality or variation on that chromosome arm. For example:

    46,XY,16 qh+

    which signifies enlargement of the secondary constriction on the long arm of chromosome 16, a common variation and unconnected with any abnormalities in phenotype. The term 'h+' signifies the presence of additional (+) heterochromatin (h).

The exact positions of abnormalities and the limits of rearranged segments of chromosomes are defined by a system in which the arms are divided up into regions numbered from the centromere outwards towards the chromosome end, bands numbered in the same manner within regions, and sub-bands within each band, again numbered from centromere outwards (Figure 11.1). For example:

   46,XX,del (1) (q21q31)

would signify that there was a deletion (*del*) in the long arm (*q*) of chromosome 1 (*1*) and the deleted segment was between band number 1 in region 2 (*21*) and band number 1 in region 3 (*31*).

The same abnormality could be defined in another, more detailed, fashion by specifying what was actually left of the chromosome rather than what had been deleted and using the convention of a double colon (::) to signify a breakage and reunion event such as would be involved in generating a deletion. The expression 'ter' signifies the end of the chromosome arm.

   46,XX, del(1) (pter ➔ q21 :: q31 ➔ qter)

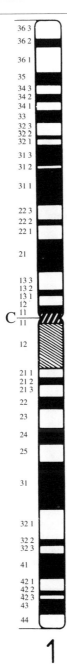

Figure 11.1 Giemsa G-banding pattern for human chromosome no. 1 at the level of resolution that shows about 500 bands per haploid chromosome set. Note how the bands are numbered outwards from the centromere (C) in each arm. The first digit in the band number designates the region, as for example in 2̲23. The second digit designates the subregion, 22̲3. The third digit designates the band, 22̲3̲.

Three more examples follow, together with a glossary of terms and conventions that will help towards an understanding of this very important 'language' of human karyology (Table 11.1).

A duplication of sub-band number 2 in band 2 of region 2 on the long arm of chromosome 10 would be defined as:

46,XY, dup (10) (q22.2)

Table 11.1

Abbreviations	Description
ace	Acentric fragment
b	Break
cen	Centromere
cs	Chromosome
ct	Chromatid
del	Deletion
der	Derivative chromosome
dup	Duplication
f	Fragment
fra	Fragile site
g	Gap
h	Secondary constriction (heterochromatin)
i	Isochromosome (one chromosome with two identical arms)
ins	Insertion
inv	Inversion
mat	Maternal origin
mos	Mosaic
p	Short arm
pat	Paternal origin
q	Long arm
rcp	Reciprocal
rea	Rearrangement
rob	Robertsonian translocation (two acrocentrics to a metacentric)
sce	Sister chromatid exchange
t	Translocation
ter	End of chromosome (terminus)
var	Variable chromosome region

Symbols		Description
Arrow	→	From to
Colon	:	Break
Double colon	::	Break and reunion
Minus	−	Loss
Plus	+	Gain
Semicolon	;	Separates chromosomes and chromosome regions in rearrangements involving more than one chromosome

An inversion of the pericentric region between p21 and q31 on chromosome 2 would be defined as:

46,XY, inv (2) (p21q31)

or

46,XY, inv (2) (pter ⟶ p21 :: q31 ⟶ p21 :: q31⟶ qter)

The last example is a reciprocal translocation in which breaks and reunions have occurred at bands q21 on chromosome 2 and q31 on chromosome 5 and the segments distal to these bands have been exchanged between the two chromosomes. Notice here how the chromosome with the lowest number (chromosome 2) is designated first.

46, XY, t (2;5) (2pter ⟶ 2q21 :: 5q31 ⟶ 5qter ; 5pter ⟶ 5q31 :: 2q21 ⟶ 2qter)

Note here how the semicolon (;) is used to separate the two chromosomes that are involved in this structural rearrangement.

THE PROBLEM OF ANEUPLOIDY

Undoubtedly the launching pad for the science of human cytogenetics was the discovery by Lejeune in 1959 of the correlation between the presence of an extra chromosome and Down's syndrome. Today there are at least 50 clinical conditions that are associated with chromosomal abnormalities. This not to say that each or indeed any of them is *caused* by the alteration in karyotype. The question of cause is very poorly understood, as we shall see later. Correlation, however, is of immense practical importance, since it allows us to make valid predictions and it serves as a sound base from which to offer advice to prospective parents with regard to the risks of pregnancy and recommendations for termination of pregnancy. It also allows clinicians to anticipate ante- and post-natal problems and make informed recommendations regarding post-natal care, quality of life and life expectancy. In what follows, we shall consider just two of the most common chromosomal disorders, Down's syndrome (trisomy for chromosome 21) and Turner's syndrome (monosomy for the X; absence of either a second X or a Y). In doing so, we shall, in fact, be looking at a significant component of the work of an average clinical laboratory. Turner's, in all its wide-ranging degrees of severity, is one of the commonest of chromosomal problems that are identified in children and young adults. True, there is not much we can do about it (endocrinologists may be able to improve growth and normalize certain morphological features by hormone treatments), but at least if we know that a person is lacking all or part of an X chromosome, then it is bound to be better than fruitlessly and perhaps even harmfully searching around for other non-existent causes for their disabilities. Down's represents the major component of all antenatal diagnosis of chromosomal abnormalities because it shows a higher incidence in older women – and women over 35 are specifically and preferentially advised to request an antenatal karyotype analysis of their unborn children. In most cases where trisomy 21 is detected in an unborn child before the 20th week of pregnancy, termination of pregnancy is offered.

Down's and Turner's represent examples of aneuploidy. What is the overall incidence of aneuploidy in humans? Estimates, some of which are based on questionable evidence, suggest that between 5 and 50% of all human conceptuses are aneuploid. One of the better and more 'scientific' estimates comes from the measurement of aneuploidy in human sperm used for *in vitro* fertilization of hamster ova. Of these, about 2% are aneuploid and a further 7–8% show structurally abnormal chromosomes. The proportion of human ova that are aneuploid has been estimated, on the basis of experience from *in vitro* fertilization programmes, to be in the region of 20–25%. This is not seriously out of line with clinical records, which show that 33% of conceptuses, based on evidence of human chorionic gonadotrophin in mothers' urine, do not survive for long after implantation. Add to this the following observations and it becomes possible to put together a reasonable view of the incidence and effects of aneuploidy in an average human population:

- nearly all embryos that are monosomic for any of the autosomes die in the first 3–5 weeks;
- 10% of all therapeutically aborted fetuses are chromosomally abnormal;
- 50–60% of all early spontaneous miscarriages between 8 and 13 weeks involve fetuses that are chromosomally abnormal;
- of all aborted fetuses with chromosomal abnormalities: 50% have autosomal trisomies, 25% are polyploid, 20% are 45, X;
- after 28 weeks of gestation, only 0.6% of live-born embryos are chromosomally abnormal;
- a significant proportion of mentally handicapped people have chromosomal abnormalities, and amongst these X and Y chromosome abnormalities are not uncommon;
- there is a higher than average frequency of sex chromosome aneuploidy (XXY, XYY, XXYY) among individuals imprisoned for certain forms of antisocial deviant behaviour.

The problem of aneuploidy can be divided into two unrelated parts, cause and effect. Cause is very much a question of cytology and particularly the cytology of cell division, meiosis, disjunction and non-disjunction and the circumstances that increase the incidence of non-disjunction. It really is very much to do with the chromosome, the spindle, the division process, and what we know mostly comes from studies of non-human systems.

Effect is a matter for the developmental biologist. Why, for example, does trisomy 21 correlate with a certain unique and body-wide combination of departures from normal?

TRISOMY 21 AND DOWN'S SYNDROME

The approach to understanding the link between trisomy 21 and Down's syndrome has been to examine chromosome 21, which seems entirely reasonable. If trisomy results from a non-disjunction during gametogenesis in the mother then the infant begins life with three 21s and develops abnormally: trisomy 21 may therefore be said to be the most likely **root cause** of the abnormalities. If trisomy were to result from non-disjunction in an early fetal mitosis, then our view of the problem might be a little less uncompromising: the non-disjunction and the attendant developmental abnormalities might just conceivably be joint consequences of something quite different. I mention this because of the immense importance of distinguishing between **cause** and **correlation** in relation to human biology. Trisomy 21 is a clear and simple deviation from normal. Down's syndrome is horrendously complicated. The temptation to take the easy route and assume from the outset that they represent cause and effect could be almost irresistible, yet if we were to do this with all human chromosomal changes and their related clinical syndromes we would undoubtedly make many mistakes and waste oceans of time and resources.

Chromosome 21 consists of about 60 000 kilobases of DNA, representing 1.6% of the haploid human genome. It probably includes several hundred expressed genes, but only a small number of these have actually been identified and mapped. What we know about the location of the factors that consistently correlate with the outward physical appearance and mental retardation of Down's patients comes almost entirely from either looking at the chromosomes of people who have these characteristics or carefully assessing the phenotypes of persons who have karyotypic abnormalities that involve chromosome 21. It is a wholly opportunistic approach. There is really no other way.

A map of chromosome 21 is shown in Figure 11.2. People trisomic for the long arm of chromosome 21 have Down's syndrome. This much we know from situations where trisomy 21 results from the possession of an **isochromosome** that consists of two chromosome 21 long arms, together with one normal and complete chromosome 21. 'Down's gene' is in the long arm. Then there are people, very rare people, who have translocations or duplications involving only a part of 21q. A careful assessment of their phenotype and the details of their chromosome 21 complement narrows the critical region down to 21q22, in effect the distal half of the long arm. A girl is found with elevated *SOD-1* and a few of the features associated with Down's. She is karyotyped as:

46, XX, dup (21) (q21 ⟶ q22.2)

A mentally handicapped but otherwise normal boy is found with:

46, XY, dup (21) (21qter ⟶ q21)

On the basis of these kinds of observations the Down's region is now considered to be on 21q and between 21.1 and 22.2, including the *SOD-1* locus at 22.1.

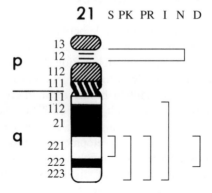

Figure 11.2 Regional map of chromosome 21 showing the regions that are known to include each of five different factors and the consensus region for Down's syndrome: N, nucleolus organizer; S, superoxide dismutase 1 (*SOD-1*) gene; PR, phosphoribosylglycinamide synthetase; I, interferon α/β receptor; PK, phosphofructokinase; D, consensus region for Down's syndrome.

However, a cautionary note has to be added here. Down's syndrome includes abnormal features of physique, mental retardation, congenital heart disease, abnormalities in the immune system, increased leukaemogenesis, heightened radiation sensitivity and presenile dementia. There has simply not been enough clinical material available to draw conclusions concerning the association of all these characteristics with a specific segment of chromosome 21. Therefore it cannot at present be said that *only* genes located in 21q22 are responsible for the specific features of the Down's syndrome phenotype when present in triplicate.

Identifying the chromosome region that correlates with Down's is one thing. Explaining the phenotype is quite a different matter. There are basically three ways to deal with this kind of problem. The first is a shotgun approach. Identify as many genes as possible within the 21q22 region that exhibit a dosage effect in Down's patients and consider how elevated levels of expression of each of these genes might conceivably be associated with the development of Down's. It is a most unsatisfactory approach since the chances of being able to explain the complexity of Down's by looking at overexpression of just a few enzymes are remote. Nevertheless, such things have to be done.

The second approach is a cellular one and its value is plain to see. Consider the primary embryonic layers in early vertebrate development: ectoderm, mesoderm, endoderm. These primary layers are established during gastrulation in the earliest stages of embryogenesis. The differentiation of these layers and the subsequent differentiation of cells within each layer depends almost entirely on three things, timing, recognition and adhesiveness, in other words, the cell cycle and the cell surface. A slight shift in any of these factors at an early stage of development could have long range knock-on consequences that will be manifest throughout the whole of the adult form. For example, correct positioning and distribution of endoderm is essential for normal cardiovascular differentiation. Suppose we introduce a slight shift in cellular behaviour during gastrulation. This will result in abnormal distribution of endoderm, which in turn will affect cardiovascular differentiation and possibly also the relative distributions of ectoderm and mesoderm in such a way as to interfere with normal development of the nervous system, and, hey presto, we have the beginnings of something on the scale of a Down's syndrome, when all we did at the outset was to slow down cell division for a few cycles, alter the properties of one cell-surface macromolecule or briefly delay the movement of one particular group of cells.

The third approach is a comparative one. Can we establish an animal model? Can we make a Down's mouse? Perhaps first of all, can we identify a 'Down's-type' linkage group anywhere in the mouse genome? Of the small number of gene loci that have been mapped to the long arm of human chromosome 21, three are present on mouse chromosome 16 and two of them lie together in a small defined region. This strongly suggests that a homologous chromosome region of significant size has been evolutionarily conserved between humans and mice. Is there a Down's syndrome region also in this common segment? Perhaps. The question, in any event, is too naive to be particularly useful, since a Down's mouse is certainly not going to look or behave like a Down's human and, besides, just making a trisomic 16 mouse could have all kinds of unrelated consequences. What we do know is that a 'Down's mouse' (trisomic 16) survives, has extensive morphological defects, including cardiovascular malformation, has reduced brain weight and is relatively short-lived. In this sense it is undoubtedly worth investigating in our quest for understanding of the human Down's syndrome.

It is of interest to note that there are recorded examples of chimpanzee and orang-utan with trisomy 22 (homoeologous to human chromosome 21, see Chapter 2) which closely resemble the clinical picture of human Down's syndrome.

MONOSOMY X

More has been written about monosomy X than on any other topic in human cytogenetics. Much of the useful research on this matter has been carried out on mice, and indeed throughout this particular story there is excellent interplay between information derived from mouse and human, altogether producing harder scientific evidence than is available for trisomy 21.

There is no evidence for X chromosome inactivation prior to the blastocyst stage (40–50 cells, 3–5 days after coitus in mouse). There is evidence from studies of dosage effects in respect of several X-linked enzymes that both X chromosomes in female embryos are active up to blastocyst. One X chromosome becomes inactivated in the trophoblast cells 4 days after coitus in mouse (see also Chapter 6). One or two days later, one of the X chromosomes of the embryo cells becomes inactivated.

At the onset of meiosis in the young ovary, X inactivation is reversed and the XX:XY or XX:XO dosage ratio progressively increases from 1 to 2. This reactivation takes place at around 13 weeks in the ovarian cells of the human female fetus.

Some X chromosome loci are not inactivated. Notably, region (Xp22.3pter), including the Xg blood group locus and the steroid sulphatase locus, remains partially active (Figure 11.3) with XX/XY dosage ratios of between 1.4 and 1.7.

It is very important to remember in relation to the effects of X inactivation and the X chromosome deficiency in a 45,XO person that they are *not* deficient when compared with a normal male (46,XY). Therefore X chromosome monosomy is not the same as an autosomal monosomy, which is always lethal.

On the other hand, since the X is not inactive from the initiation of female meiosis through to ovulation in the adult female, defects in oogenesis in XO females may justifiably be considered as consequences of X chromosome monosomy. XO females are born without functional ovaries, although they do produce some eggs during embryonic life which degenerate just prior to birth. XO mice produce eggs, but their fertility is relatively short-lived and runs out after a few months.

There are therefore three basic questions that serve as research foci in relation to the XO syndrome. First, are the abnormalities connected with deficient levels of expression of those genes on the X chromosomes that are never inactivated? Second, how critical is the activity of the two X chromosomes during the very early stages of development, when they are normally both active? Finally, is activity of both X chromosomes in the developing ovarian egg critical for normal growth of that egg, for overall development of the possessor of that egg and/or for development of the embryo into which that egg may subsequently grow? It is plain to see that all these and related questions can be investigated in mice and other animal systems much more easily than in humans, and just as profitably.

THE CLINICAL CYTOGENETICS LABORATORY

The day-to-day work of a typical clinical cytogenetics diagnostic laboratory in the UK might be said to fall into three major categories. The first is Down's, the second sex chromosome abnormalities and the third is all the rest. Down's is regarded as serious and generally justifies an offer of termination of pregnancy if detected antenatally. Sex chromosome abnormalities and their phenotypic consequences are less predictable, less serious and arguable as justification for terminations of pregnancies. The rest of the wide range of chromosomal ab-

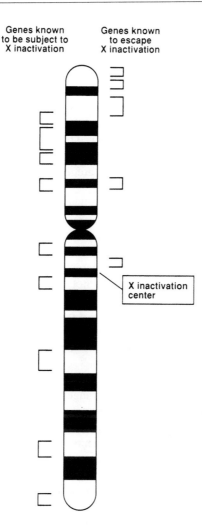

Genes known
to be subject to
X inactivation

Genes known
to escape
X inactivation

X inactivation
center

Figure 11.3 Map of the human X chromosome (based on Thompson *et al.*, 1991). The positions of some of the genes known to be subject to X inactivation are shown by brackets on the left of the map. The positions of others known to be unaffected by X inactivation are shown on the right. The non-inactivated region is at the extreme distal end of the p (short) arm, position 22.3.

normalities are, for the most part, either relatively harmless, necessitating counselling but little else, or extremely harmful and certain to lead to death of the fetus. Those chromosomal abnormalities that generate viable but severely handicapped children thankfully constitute a relatively small minority of cases and can only be dealt with through the most careful and compassionate co-operation between parents, doctors and genetic counsellors.

In the UK, most of the work of a clinical laboratory comes from hospital consultants to whom a patient has been referred by his or her family doctor. Referrals fall into five major categories.

1. *Antenatal clinics.* In the UK there has been, until recently, a strong selection for mothers aged over 35, and in these the risk of occurrence of Down's is considerably higher than average. These clinics provide samples of chorionic villus tissue or amniotic fluid containing fetal cells. Long-term tissue culture (2–3 weeks) is then involved followed by the examination of a relatively small number of mitotic figures with two principal and overriding questions in mind – how many chromosomes are there, and, if there are 47, is the extra one a small acrocentric?

   In 1992 in a medium-sized English county there may be around 12 000 births a year, of which 4–5% are screened for chromosomal abnormalities, but the vast majority of these will have been preselected on account of maternal age or the results of other tests that have been carried out in consideration of maternal age or parental family genetic background.

2. *Neonatal clinics.* The baby does not look or behave quite right. Is there anything abnormal with its chromosomes? If its abnormalities are seriously life-threatening, do we try to save it or not. If it has a major chromosomal abnormality it is most unlikely to survive for long and major medical or surgical intervention may accomplish nothing and be distressful to child and parents.

3. *Retarded early development.* Is there anything wrong with the chromosomes? Nearly all autosomal defects, rearrangements or imbalances have associated mental retardation. There is little or nothing that can be done about it, but it helps to know and understand, and such information is valuable in planning treatment and management of patients.

4. *Puberty.* Something is wrong. The same questions arise as in category 3.
   In categories 2, 3 and 4, *both* parents have to be karyotyped if a chromosomal structural rearrangement is suspected.

5. *Infertility and miscarriage.* Normal adult carriers of chromosomal rearrangements may produce chromosomally unbalanced or deficient gametes.

Particularly in relation to programmes of *in vitro* fertilization using either normal partners or anonymous donors, all participating donor males and females are prescreened for chromosomal abnormalities. A typical case history might be as follows:

A woman of 36 goes to her family doctor and is diagnosed as pregnant. She is referred to a consultant obstetrician in her local hospital, who advises her to undergo a test that will involve sampling fetal tissue. The samples are passed to the regional cytogenetics laboratory with meticulous attention paid to identity.

Three weeks later a result is passed back to the obstetrician. A transmissible chromosomal abnormality has been detected. Where did the abnormality come from? The obstetrician calls for blood samples from the parents. The father shows evidence of the abnormality. His chromosome preparations are subjected to detailed investigation and the precise nature of his chromosomal rearrangement is determined employing several different chromosome banding techniques. A report is passed to the obstetrician who then immediately refers the couple for genetic counselling. After genetic counselling, the parents then decide whether or not to opt for termination of the pregnancy. At a later genetic counselling, the consultant geneticist and the clinical cytogeneticist assess the risk factors in relation to the father and discuss these with both parents with a view to informing them of the likelihood and consequences of the father transmitting harmful chromosomal defects to future children. At this time, other family members who may possibly carry the chromosomal abnormality are traced and karyotyped.

The range of technologies available to clinical cytogeneticists is wide, and selection of the right approach is perhaps the most skilled and critical matter in the entire job. Most laboratories now employ techniques of tissue culture that lead to some degree of synchronization of cells in culture, such that it becomes possible to harvest cells that are specifically in late prophase or early metaphase of mitosis. This produces longer chromosomes and the potential for better resolution of bands. The standard technique is G-banding, but other methods have special and extremely valuable applications.

Q-banding, for example, is a quick method for distinguishing between an extra Y chromosome and an extra 21. Both chromosomes are morphologically similar and sometimes indistinguishable after conventional staining, but the long arm of the Y fluoresces brilliantly with quinacrine (Figure 2.2). C-banding can be extremely useful in searching for pericentric inversions or dicentric chromosomes. A dicentric X, for example, is not uncommon and will usually be an iXq (Figure 11.4), two long arms, two centromeres and a little bit of the combined short arms between the centromeres, a situation that is immediately decipherable with a combination of G- and C-banding. More modern techniques of *in situ* nucleic acid hybridization and 'chromosome painting' are of special use for identifying the component parts of chromosomes that are involved in translocations, since they enable us to identify individual chromosomes or parts of them by colour alone.

It is easy to see, then, that the skill of a clinical cytogeneticist ranges far beyond the relatively routine tasks of cell culturing and chromosome identification. It is undoubtedly important to be able to apply the technologies and interpret the results correctly, but the detective component of the field is

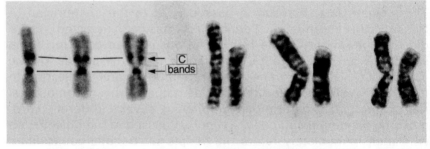

Figure 11.4 Abnormal X chromosomes from cells of a female person after C-banding and G-banding. In the three G-banded examples on the right, the abnormal X is to the left of the normal one from the same cell. In the C-banded preparations on the left, only the abnormal Xs are shown: each of these shows two centromere regions (C-bands) – a dicentric condition – with a short interstitial centromeric region between them, confirming that these are dicentric chromosomes. That they are isochromosomes for the q arm of the X is evident from the G-band patterns. (Reproduced with permission from Howell, R.T., Roberts, S.H. and Beard, R.J., *J. Med. Genet.* 13, 496–500, 1976.)

perhaps most important of all. Human cytogeneticists have to be keen-eyed, critical, clever and patient.

The job of analysing human karyotypes can be tedious. Occupational hazards range from eyestrain to inflammation of the tendons in the wrist from continuous small movements of the controls on the microscope. Can the task be automated? Partially automated systems for human karyotype analysis have been designed and are currently in use in many countries. The more successful of these is a combination of a modern high-resolution light microscope with a motor-driven mechanical stage and video camera linked to a scanner and computer. The system is designed to scan systematically across the whole area of a microscope slide preparation at low magnification, recording images as it goes, recognizing human chromosome spreads, registering the coordinates for each spread. The entire scanning process for one preparation covering perhaps 2 cm² takes less than a minute. The computer then sorts the metaphases out into a rank order of quality based on completeness and absence of overlaps between chromosomes. It can then be asked to sort the chromosomes out into a karyogram, longest to shortest and paired according to size, centromere position and banding pattern. Only at this stage is it necessary for the human operator to intervene with judgement regarding chromosomal rearrangements. The systems are expensive, but they can be justified if they are kept in full-time use for fast processing of very large numbers of samples. Naturally, such systems are useless unless they are supplied with good-quality microscope preparations, and these still depend on the individual skills of laboratory staff, each with their own particular brand of witchcraft for the consistent production of excellent human mitotic spreads.

RECOMMENDED FURTHER READING

Epstein, C.J. (1986) *The Consequences of Chromosome Imbalance*. The University Press, Cambridge.

Howell, R.T., Roberts, S.H. and Beard, R.J. (1976) Dicentric X isochromosomes in man. *J. Med. Genet.* **13**, 496–500.

ISCN (1985) An International System for Human Cytogenetics Nomenclature: Birth Defects. Original Article Series, 1985, Vol. 21, No. 1. March of Dimes Birth Defects Foundation, New York.

Smith, G.F. (Ed.) (1985) *Molecular Structure of the Number 21 Chromosome and Down Syndrome*. The New York Academy of Sciences, New York.

Thompson, M.W., McInnes, R.R. and Willard, H.F. (1991) *Genetics in Medicine*, 5th edn. W.B. Saunders, Philadelphia.

Verma, R.S. and Babu, A. (1989) *Human Chromosomes: Manual of Basic Techniques*. Pergamon Press, Oxford.

RECOMMENDED FOR READING

Outlook for 2001

The two decades leading up to 1993 are simply packed with progress in cyto-genetics and they will be a very hard act to follow through into the twenty-first century.

Consider the scene four decades ago. Phase-contrast microscopy was in the development stage. The first transmission electron microscopes were beginning to appear in a few scattered research laboratories and they were primitive and unreliable. The unique genetic role of DNA had just been discovered. The human chromosome number was 48 and no-one had ever seen a whole human karyotype. Computers did not exist, except for the ingenious mechanical ones that were used for bomb sighting in the 1939–45 war. Cytology was all in black and white and although the quality of the picture was often superb, our understanding of the image was severely limited. Nevertheless, the goals were clear.

First, we needed better resolution. Second, we were sure that chromosomes were linear and that each represented a single enormously long DNA double helix, so the search was on for visible linear differentiation. We had a right to expect it, since it was so obviously present in the giant polytene chromosomes that had been known and studied for more than 30 years previously. Third, we knew that genes were distributed along chromosomes and we wanted a method for finding them.

It all happened quite quickly. Resolution improved at the speed it took to develop light and electron microscopes, the latter being giving an enormous boost with the introduction of micro-electronic circuitry. We have seen in Chapter 2 of this book how chromosome banding was developed and how C-banding was very largely a spin-off from early successful attempts at *in situ* nucleic acid hybridization. Most of this happened in a heady period of intensely competitive science in the decade following 1965, and it included the discovery of restriction endonucleases in 1969 quickly followed by the invention of recombinant DNA technology for cloning of specific polynucleotides. By 1975 all the shapes were there, recorded meticulously by early observers and classical cytogeneticists. Now they had to be explained, and at the time there seemed to be a massive surfeit of tools with which to do the job. It was at about this time, in the early 1970s, that an eminent American cell biologist remarked that 'A young biologist today really has no choice but to become a molecular biologist'.

Now we have reached 1993. We have unlocked linear differentiation. Our ability to locate genes on chromosomes is limited only by the need to isolate and label the sequences. We have improved resolving power very nearly to the theoretical limits in light microscopy and electron microscopists have long since been operating at the molecular level. We have unravelled the chromosome and, through a crafty blend of microscopy and molecular biology, we now understand the three levels of organization that are represented in the condensed metaphase condition. We are rapidly learning about centromeres and kinetochores. We know a lot about telomeres, how they seal chromosome ends and render them non-interactive, and how they facilitate the completion of DNA replication at the end of the duplex. In some cases we have actually mapped the gene sequences from end to end on specific chromosomes and we are well on the way to generating entire chromosome nucleotide sequence maps (Chumakov *et al.* 1992). We have entered the era of the 'domain', central domains, surface domains, centromere domains, pairing domains, coding domains, specific notional aspects of a chromosome that are functionally important or intimately involved in events that we are able to define – but about which we know practically nothing! The truth in 1993 is that after nearly 100 years of chromosome research we have almost completed the basic foundations of an understanding in this field. The artistic and colourful illustrations that decorate the pages of modern undergraduate textbooks in cell biology are drawn with confidence and memorized with ease. So long as the chromosome remains still and conforms, we are

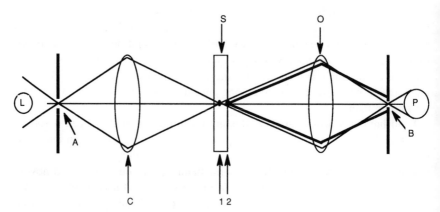

Figure 12.1 Optical path in a simple confocal system. The condenser lens (C) forms an image of the first pinhole (A) onto level 1 in the specimen (S). The objective lens (O) forms an image of specimen level 1 at the second, exit, pinhole (B), which is confocal with (A) and with specimen level 1. The image is picked up by the camera at (P) on the distal side of exit pinhole (B). Light from another focal level (2) in the specimen is not focused at (B) and most of it will not pass through pinhole (B) (the thick lines in the ray diagram).

comfortable. As soon as it begins to move, however, or changes its shape, our confidence evaporates, for it has to be admitted that at this point in time we simply do not know what is going on.

What can we do today that we were unable to do 10 years ago? The answer is: not a lot, but we can do it much faster. A prime example is the development of the confocal microscope. Perhaps the most astonishing thing about confocal microscopy is that in 1988 it generated instant excitement and immediately sold itself as a major new tool for biologists: yet the principle of confocal microscopy was discovered and patented and its potential fully appreciated as early as 1957.

The basis of the technique is as follows. When we look at something down a microscope we use the fine focus knob to adjust the focus so that we are looking at an image in one focal plane of our specimen. If the specimen is thin, less than 1 or 2 μm, then the image is quite clear and contrasty. If the specimen is thick, more than about 5 μm, then our image becomes less clear, and still less clear as we examine thicker and thicker specimens. The image is deteriorating because the scattered light from the out-of-focus images of material above and below our plane of focus is fogging and obscuring the image of the thing we are trying to look at. In 1957, Marvin Minsky, a postdoctoral fellow at Harvard University, thought of a way of overcoming this problem.

In Minksy's microscope (refer to Figure 12.1), the condenser lens is replaced by a lens similar to the objective (C). The field of illumination is limited by a pinhole (A) positioned on the optical axis of the microscope. A reduced image of this pinhole is projected onto the specimen (S) by the condenser (C). The field of view is reduced by a second pinhole (B) in the image plane, placed confocally to the illuminated spot in the specimen and to the first pinhole (A).

The specimen is scanned through a point of light by moving it in a raster pattern by some system that generates controlled movement of the microscope stage. Variation in the density of the image during the scan is picked up by a photomultiplier tube, digitized and transmitted to form an image on a television screen. All very straightforward, but what is so new and wonderful?

Examine carefully the diagram in Figure 12.1 and it should be apparent that light that is scattered from parts of the specimen above or below the plane of focus (thick lines in Figure 12.1) is rejected from the optical system: it does not get through the imaging pinhole (B). Consequently, the image of whatever is in the objective's plane of focus will be clear and unaffected by the out-of-focus images of material at higher or lower focal planes. What could be simpler?

The confocal microscope uses a coherent laser light source to increase resolution and brightness and it requires that the specimen be stained with a fluorescent dye so as to produce bright images against a black background. It then allows us to produce optical transverse sections though our thick specimen, each one being a crisp image of whatever is in focus at a particular level.

Before confocal microscopy, the only way to visualize biological microstructure in three dimensions was to histologically fix a specimen, embed it in wax or plastic, cut it into a continuous series of thin slices (sections), mount these on microscope slides in a numbered sequence, stain them, photograph them and then generate a three-dimensional image by stacking the pictures on top of one another in the correct order. A relatively recent example of this kind of approach can be found in Diter Von Wettstein's (1984) studies of the orientation of synaptonemal complexes in *Bombyx* spermatocytes. It takes little imagination to sense how much time and work is involved in a process of this kind.

In the author's most recent experience a three-dimensional reconstruction of woodpigeon's lampbrush chromosomes arranged within an oocyte nucleus one-fifth of a millimetre in diameter was obtained, by confocal microscopy, less than 2 hours after the bird had been killed.

Naturally, the success of the confocal microscope is not entirely due to the principle discovered by Minsky. The full potential of his invention could never have been realized in 1957, because there were no computers. It is one thing to obtain a clear image, still another thing to generate a stack of images, quite another thing to integrate these images into a three-dimensional reconstruction. The last simply cannot be done without a computer. But once you have your computer linked to your confocal system, then you can do all kinds of things, magnify and demagnify images, scan in the z-axis, measure depth, thickness, density, produce complex image enhancement effects – all of this away from the microscope, since once the images are generated and stored on disk, the microscope is no longer needed and can instantly be made available to the next eager investigator in the queue.

Confocal microscopy is just one of the refinements that have speeded up cytological investigation in recent years. The other clear example is the use of biotin as a label for molecular probes used for *in situ* nucleic acid hybridization and the detection of the biotinylated probe with an antibody to biotin labelled with a fluorescent dye. The technique has acquired the acronym FISH (fluorescence *in situ* hybridization). Before FISH was first introduced by Langer and Ward in 1981 (see Korenberg *et al.* 1992, Trask 1991) all *in situ* hybridization was carried out with radiolabelled probes that were detected by autoradiography. Microscope slides bearing radiolabelled specimens were coated with nuclear track emulsion, sensitive to beta-emissions from the radioisotope. They were left in the dark for the emulsion to 'expose' in exactly the same sense as with a photographic film. Then, hours, days, weeks and sometimes even months later the slides were 'developed' and the autoradiograph could be examined and the location of the *in situ* hybrid determined. With FISH, the entire operation takes about 2 hours.

In the midst of all this modern high technology, most of which, incidentally, is very expensive, the exploitation of animal and plant diversity will remain for ever by far the most powerful scientific tool that we have in our box. It is a multi-purpose tool. Diversity represents a living record of nature's successes and failures within the framework of evolution and natural selection. Show biologists something they have never seen before in an animal and the first question they will ask is whether the same thing or a modified form of it is present in other animals.

If they find a modified form, then they will search further afield for other forms on the principle that by looking at nature's variations they will be able to sort out the important features of the new phenomenon from the trivia and get closer to an understanding of both form and function.

The other sense in which diversity is immensely important is that for each and every biological problem, for each hypothesis, for each idea, there will be one particular type of organism that is better suited than any others for solving or testing it. The pages of science are full of examples of this principle. The Indian muntjac is best for investigating the structure of kinetochores because two of its chromosomes have exceptionally big ones. *Notophthalmus* has been a favourite for studies of lampbrush chromosomes, because its are amongst the largest that are to be found. David Shaw and his associates in Australia chose the grasshopper *Caledia captiva* to test the idea of the 'coadapted' genome because within that species there are innumerable small isolated populations that have independently evolved small variations in their karyotypes, whilst remaining fully capable of producing 'interpopulation' F1 and F2 backcross hybrids. Horses and donkeys are ideal for testing random X inactivation because they can produce F1 hybrids (mules) and their X chromosomes are completely different in size and shape. *Bombyx*, the silk moth, is ideal for investigations of synapsis, synaptonemal complexes and recombination nodules because it has a nice orderly synaptic sequence in the male and it forms no chiasmata in the females (Von Wettstein 1984).

Sometimes, the investigator is just plain lucky and gets a bonus in addition to choosing an organism that is right for the job. Two of the best examples of lucky investigators are Gregor Mendel and Herbert Taylor. Mendel chose two sets of characters in *Pisum sativum* that just happened to be on different chromosomes, so his results were uncomplicated by linkage. At the time, of course, he did not even know that chromosomes existed. Taylor used *Vicia faba*, the broad bean, for his experiment to test the idea of semiconservative replication of chromosomes (Figure 7.1) and, lucky for him, there were no sister chromatid exchanges; otherwise he would have had the greatest of difficulty in explaining what he saw.

What are likely to be the priority areas for future research in cytogenetics? The topics are easy to list. Meiosis remains full of unknowns and most of them are quite formidable. What are recombination nodules? How does the synaptonemal complex work? How are interlocks resolved? What brings homologous chromosomes into pairing association with one another? Have all the features of meiosis evolved by a process of natural selection, and, if so, what are the selective advantages of each? Chromosome disjunction must be high on our list of priorities. We are, to be sure, well advanced with our understanding of centromeres, kinetochores, microtubules and chromosome movement – an altogether fascinating area of modern cell science – but there is still a long way to go before we understand well enough to explain the countless diversities of disjunctional behaviour and spindle mechanics.

Imprinting and sex will remain high-priority areas. How does a chromosome 'know' it has a paternal origin? Will we ever be able to understand the molecular basis of events such as those that accompany gametogenesis in males of *Sciara coprophila* (see Chapter 6 and Figure 6.3)? What is the molecular basis of chromosome inactivation? What is the molecular basis of sex determination, for here we have the prime example of a limited number of genes localized in one small region of a specific chromosome operating as a major developmental switch mechanism?

Nucleus and karyotype have become specially important in recent years, perhaps because computers, confocal systems, interest in DNA–protein interactions and the development of chromosome 'painting' and whole-genome *in situ* hybridization (GISH) have enabled us to consider whether there is more to an interphase nucleus than just a membrane-bounded compartment occupied by jumble of threads. In the past we have been so preoccupied with the orderly events of cell division that the interphase nucleus has seemed unimportant. The increasing volume of evidence showing that what goes on at cell division is strongly influenced by events in interphase compels us to shift our attention and ask about the relative positioning of chromosomes in relation to one another and to the things that happen to them earlier or later in the cell cycle. It is a field that was pioneered by two well-known British plant cytogeneticists, Michael Bennett and J.S. (Pat) Heslop Harrison, and their tenacity and skill has paid off handsomely. Nuclei are now seen as dynamic structures with a design and structure that is both flexible and versatile, and there is absolutely no doubt that this is going to be an exceedingly active field of research in years to come.

To many cytogeneticists, karyotypes have become more interesting than individual chromosomes. An organism's karyotype is seen as a species-specific character that has evolved through natural selection. It is less likely to be of adaptive significance to the individual than to the population. We can do almost anything with the chromosome set of *Drosophila melanogaster* in the sense of rearranging it or chopping it up and we still get a *D. melanogaster*. On the other hand, we cannot mix two karyotypes that have evolved separately, even though they contain the same genes and derive from the same 'species', and expect to get viable offspring. The 'coadapted genome' is something we do not yet understand and it will doubtless occupy many cytogeneticists in the years ahead.

Of course the impact of cytogenetics is already immense, though more so in relation to plants than animals at this point in time. More than half the British wheat crop, for example, and vast areas of crop agriculture in other parts of the world are the result of cytogenetic manipulation – the directed incorporation of alien chromosome segments from different species into wheat. Chromosome engineering, currently the most important area of genetic engineering, is likely to be even more widely applied in agriculture and perhaps eventually also in medicine, if the ethical problems can be resolved: but then, the ethics of the twenty-first century are sure to be very different from those of 1993! Transferring genes has to be regarded as the high-impact biotechnology of the twenty-first century, and in this sense an expanded understanding of chromosome recombination, genome behaviour in hybrids, gene expression and chromosome evolution is likely to be hotly in demand.

Cytology and cytogenetics are fields of biology where major advances can be expected quite soon, mostly based on the time-honoured principle that if you start out at the molecular level with a simple organism whatever you find out is likely to provide some pretty strong clues about what happens in more complex situations. With regard to the cell cycle, for example, we are already in the midst of an exciting and productive interplay between studies of centromeres in yeast and larger eukaryotic cells, for in this field researchers are taking advantage of the superior cytology of the big cells, the superior genetics of yeast and the common ground of biochemistry and molecular biology.

Nevertheless, cytogenetics remains a field where it is still possible for one person on their own, working with an unusual organism, to make a discovery that will put them into the history books of science. The unknowns are tough tests of human ingenuity and experimental skill, but we should not let that deter us from having a go. When we reach 2093, cell biologists will be as familiar with the workings of living cells as a skilled mechanic is with the average car engine. But then, few people who read this book will be alive in 2093 and those who are will not be looking at chromosomes. Our problems are today's problems, and we had best get on with solving them and be thankful that we work with chromosomes and not with motor cars.

RECOMMENDED FURTHER READING

Chumakov, I. *et al.* (more than 20 co-authors) (1992) Continuum of overlapping clones spanning the entire human chromosome 21q. *Nature* **359**, 380–387.

Heslop Harrison, J.S. (1993). Nuclear architecture in plants. *Current Opinion in Genetics and Development,* in press.

Korenberg, J.R., Yang-Feng, T., Schreck, R. and Chen, X.N. (1992). Using fluorescence in situ hybridization (FISH) in genome mapping. *Trends Biotechnol.* **10**, 27–32.

Koshland, D. (1992). Mitotic motors. *Current Biology* **2**, 569–571.

Minsky, M. (1988). Memoir on inventing the confocal scanning microscope. *Scanning* **10**, 128–138.

Pawley, J. 1990. *Biological Confocal Microscopy*. Plenum Publishing Corporation, New York.

Rattner, J.B. (1992). Integrating chromosome structure with function. *Chromosoma (Berl.)* **101**, 259–264.

Trask, B.J. (1991). Fluorescence in situ hybridization: applications in cytogenetics and gene mapping. *Trends Genet.* **7**, 149–154.

Von Wettstein, D. (1984) The synaptonemal complex and genetic segregation. In *Controlling Events in Meiosis*. C.W. Evans and H.G. Dickinson (Eds.). Symposium of The Society for Experimental Biology XXXVIII. The Company of Biologists, Cambridge, pp. 195–232.

APPENDIX   A selection of laboratory practical protocols

INTRODUCTION

Each of the protocols presented here has been tried and tested over a period of 22 years in teaching laboratories in Europe and the United States. They have been operated successfully with undergraduate classes of between 20 and 70 students and an instructor–student ratio of around 1:10. Each protocol is best suited to a timetable that allows one full day (minimum 5 hours) in the laboratory followed later by a second session of about 3 hours. The particular timing will, of course, have to be decided by the course organizer in relation to his or her resources and constraints.

All the protocols have been consistently successful, and instructors can be confident that each will provide good experience and enjoyment for students under any circumstances short of total disaster.

The protocols assume only that the students have a basic grounding in the use of a light microscope and that they have some basic training in the most elementary laboratory procedures. The author has always commenced this particular programme of laboratory practicals with a 2-hour session in which students, under close supervision, have to reassemble and set up for proper use the compound phase-contrast light microscope that they will use for the remainder of the course. That exercise begins with a microscope that has been dismantled into its body, stage, condenser, objectives and eyepieces, all neatly and systematically arranged at the student's workplace.

It is recommended that each exercise be accompanied by a set of objective questions, the answers to which provide the basic framework for the student's written report. Some examples of these questions are suggested at the end of each protocol.

Naturally, the author is aware that in some countries the use of human blood and the killing and dissection of animals in teaching laboratories may be prohibited by law. Nevertheless, all these protocols could, in such circumstances, be carried out beforehand and the prepared or semiprepared material could be provided for the students.

Further detailed information on every aspect of these protocols and instructions can be found in *Working with Animal Chromosomes* by H.C. Macgregor and J..M. Varley, John Wiley & Sons, Chichester, 1983 and 1988.

SOME NOTES ON LIGHT MICROSCOPY

Angular aperture of a lens (α)

This is defined as half the angle subtended by the cone of light entering a lens from a point source at the focal point of the lens and on its optical axis.

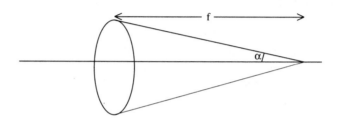

NOTE: α WILL ALWAYS BE LESS THAN 90° AND THE SINE OF α WILL ALWAYS BE LESS THAN 1.

Numerical aperture of a lens (NA)

$$NA = \mu\sin \alpha$$

where μ is the refractive index of the material between the object and the lens. For a dry lens that material will be air with $\mu = 1.0$. It follows that the NA for a dry lens will always be less than 1.

For an oil immersion lens there will be oil between object and lens, and μ will be greater than 1.0. Therefore any objective with NA of 1 or more than 1 *must* be an immersion objective.

Resolving power

The resolving power of a lens, R (expressed in the same units as used for the wavelength W) = $W/2NA$, where W is the wavelength of the light used to illuminate the specimen. This implies that the *shorter* the wavelength the *greater* will be the resolving power of the microscope. Resolving power may be defined as the capacity of a lens used with light in the visible part of the spectrum to resolve fine detail in any object. Note that the numerical aperture of an objective lens provides us with a means of estimating the *theoretical* limit of fineness of detail that we can expect to resolve with that lens.

Wavelength

Useful visible range 400–700 nm;

400 deep violet;
480 blue;
550 green (mean wavelength of white light about 550 nm);
580 yellow;
620 and up, red.

Magnification

The magnifying power of an objective is indicated by the first value, engraved in large numerals, on every objective. The total magnification of the microscope is determined by multiplying the magnification of the objective by that of the eyepiece. Thus, a 10 × eyepiece used in conjunction with a 40 × objective gives a total magnification of 400 × .

Sizes

One metre (m) = 100 centimetres
One centimetre (cm) = 10 millimetres
One millimetre (mm) = 1000 micrometres
One micrometre (mm) = 1000 nanometres
One nanometre (nm) = 10 Angstrom units (Å).

Light microscopes are unlikely to resolve detail smaller than about one quarter of a micrometre. Therefore, most measurements in light microscopy are likely to be expressed in millimetres or micrometres.

GET INTO THE HABIT OF MEASURING EVERYTHING YOU SEE.
DESCRIPTIONS ARE USELESS WITHOUT MEASUREMENTS.

Slides and coverglasses

When you use a microscope to look at a specimen mounted between slide and coverglass, the slide and coverglass become part of the optical system of the microscope. Their thickness, quality and cleanliness are therefore important. Slides are generally made of polished, high-quality glass, and are usually about

1 mm thick. British slides vary from 0.95 to 1.0 mm in thickness. American slides are usually thicker, at about 1.11 mm.

Coverglasses come in five thicknesses which are designated by numbers 0, 1, 1.5, 2 and 3. The thicknesses of these coverglasses are as follows:

0 = 0.09 mm
1 = 0.12 mm
1.5 = 0.16 mm
2 = 0.19 mm
3 = 0.25 mm

Most modern microscope objectives are calibrated for use with No. 1.5 coverglasses with thicknesses of between 0.15 and 0.17 mm. It is impossible to use No. 2 coverglasses with the oil immersion objectives, since these coverglasses are thicker than the 'working distance' of the objective, and you cannot get near enough to the object to get it into focus.

LABORATORY PRACTICAL 1: THE MITOTIC CHROMOSOME

HUMAN CHROMOSOME PREPARATIONS FROM PERIPHERAL BLOOD

Between 1955 and 1959 a group of investigators, led by Hungerford in the United States, worked on and developed a technique for culturing the white blood cells (lymphocytes) from human blood and inducing them to enter mitosis and multiply. The secret of the technique, as it was finally adopted, was the use of a substance called **phytohaemagglutinin**, a plant product which was found to stimulate division in cultured lymphocytes. The technique has gone through many modifications and minor changes, and different laboratories now have their own peculiar brands of 'witchcraft', but in its essentials it remains virtually unchanged from the original Hungerford method.

The blood is withdrawn aseptically into heparinized containers so that it does not clot. It is then mixed with a culture medium that is a complex of a wide variety of compounds needed for the survival of mammalian cells in culture. Phytohaemagglutinin is added, and the mixture is allowed to incubate under sterile conditions for about 3 days at 37°C. Colchicine is then added. This arrests dividing cells in metaphase by preventing the formation of the mitotic spindle. After a few hours, the cells are washed, fixed and spread on slides in such a way that the chromosomes of dividing cells become well separated from each other and form clearly recognizable diploid groups.

The technique described in this protocol is precisely that which is routinely used by many clinical laboratories throughout the world. The blood samples are allowed to stand for half an hour. The red cells settle to the bottom of the tube and white cells form a layer over them. A sample of plasma enriched with white cells can then be removed from the layer above the red cells. Although whole blood can be used, because we select the white cells and discard the red cells, we greatly enrich the culture with dividing cells and we should therefore see more metaphases in our final preparations.

Setting up the culture

The blood samples are obtained by venepuncture with a sterile, heparinized, tuberculin syringe with a 20-gauge needle. This operation must be carried out by a medically trained person.

The blood is transferred immediately to the culture medium in a sterile glass or plastic 20-ml culture tube with a screw-on cap. The blood can, if necessary, be stored for up to 1 week at 4°C, but in general the sooner it is used after removal from the person the better.

1. The blood is dispersed into the medium by gentle inversion of the tightly closed tube.
2. The cells are incubated for 66–72 hours at 37°C.
3. Following incubation 0.3 ml of Colcemid solution (10 μg/ml) is added to the culture and incubation continued for an additional 1.5 hours. Colcemid is a commercial preparation closely related to colchicine. It need not be sterile.

   Normally, a laboratory practical class will start at this point, the students having been provided with the tubes of blood in culture medium about 1.5 hours after the addition of the Colcemid.

THE TIMING OF STEPS 4–9 IN WHAT FOLLOWS IS CRITICAL. WEAR RUBBER GLOVES AND MOP UP ANY SPILLS WITH DISINFECTANT.

4. To harvest the cells, transfer the blood culture to the screw-cap, conical-bottomed, centrifuge tube. Centrifuge the culture in a bench centrifuge for 6 minutes at about 1000 r.p.m. The supernatant should be clear.
5. Using a plastic pipette carefully pipette off all but 0.5 ml of the supernatant. Resuspend the button of spun-down cells in the remaining 0.5 ml of supernatant; do this by flicking the tube gently. Make sure that there are no clumps of cells left unsuspended.
6. Add two drops of warm (37°C) 0.075 M KCl, the hypotonic solution. Flick the tube to mix and immediately **start timing six minutes**. Then quickly add a further 6 ml of 0.075 M KCl and leave the tube standing in a 37°C waterbath for the remainder of the 6 minutes.
7. Centrifuge the tube for 6 minutes at about 1000 r.p.m. **Total time elapsed 12 minutes**.
8. Discard all but 0.5 ml of supernatant into the waste beaker. Resuspend the cells in the remaining supernatant, making sure there are no cells stuck to the bottom of the tube.
9. Add two drops of 3:1 fixative, flick the tube to mix, add a further 4 ml of 3:1 fixative. Leave for at least 15 minutes at room temperature.
10. Centrifuge the tube for 6 minutes at about 1000 r.p.m.
11. Repeat steps 8, 9 and 10.
12. After the second fixation procedure, remove all but 0.5 ml of fixative solution. Resuspend the pellet of cells in the remaining fixative.
13. Using a siliconized glass pipette, splash one or two drops of fixed cells on a clean, chilled, wet slide from a height of about 18 inches. Blow gently over the surface until the slide is dry: this ensures better distribution of the cells. Check the concentration of metaphase spreads on the slide using phase contrast. Dilute or resuspend in a smaller volume if necessary. Then make as many slide preparations as you can but only stain about four

slides. Keep the unstained slides safe and covered to protect them from dust.

14. Giemsa staining. Place the dried preparations in a Coplin jar containing 50 ml of phosphate buffer. Add 3 ml of Giemsa stain and leave for 10 minutes. Hold the jar under the distilled water tap and flood out the stain. **It is important not to lift the slides directly out of the Giemsa**. A metallic film forms on the surface of the Giemsa solution and if you lift the slides through this they will be covered with horrible black messy crystals.

15. When you have washed all the Giemsa away, air dry the preparations as quickly as possible either by blowing on the surface of the slide or waving it about vigorously in the air. Place a drop of mounting medium, Canada Balsam, Euparal, XAM, DPX, etc., over the preparation and cover with a clean coverglass.

Place the slides on a hot plate, about 60°C, for 24 hours or more to allow the mounting medium to harden. The preparations can now be examined with normal bright-field microscopy. Special attention to the following matters is recommended.

- The shapes and absolute sizes of the largest and smallest chromosomes.
- Identification of the sex chromosomes.
- Identification of the G-group chromosomes to which chromosome 21 belongs.
- The short arms of all the D- and G-group chromosomes. These usually take the form of very small dots or satellites attached to the remainder of the chromosome by a faint tenuous region that represents the nucleolus organizer constriction.
- Associations between the short arms of D- and G-group chromosomes. In some cases several D- and G-group chromosomes may be apparently joined to one another through their short arms to form a rosette-shaped cluster.

Constructing a karyotype

The photomicrographs with which you are provided are pictures of complete chromosome sets from a variety of individuals. Some are normal. Others are abnormal and were obtained from the local Regional Cytogenetics Laboratory.

1. Note the number of the photograph.
2. Cut out each chromosome neatly. Do not lose any or duplicate any. (Two prints of each photograph are provided if there are overlapping chromosomes in the picture.)
3. Sort out the chromosomes into groups A, B, C, D, E, F, G, X and Y.
4. Arrange the chromosomes in each group in pairs in descending order of size.
5. Stick them to the card provided in lines representing the different groups.
- Can you identify any abnormalities?
- What can you say about the individual from whom the chromosomes came?

Production of idiograms

An idiogram is a diagrammatic representation of the haploid chromosome set of an organism. It follows certain conventions. Chromosomes are arranged in order of size with the largest on the left and the smallest on the right. The short arms of the chromosomes are always upwards. The sex chromosomes (X and Y or W and Z) are always placed at the end of the line. Each chromosome is represented according to its relative length and centromere index. The idiogram is always accompanied by a list of data showing the relative lengths and centromere indices. Some investigators arrange the chromosomes with their centromeres aligned. Others arrange them as in a histogram, without regard to alignment of the centromeres.

For a discussion of the conventions and methods for constructing karyotypes and idiograms see *Working with Animal Chromosomes* by H.C. Macgregor and J.M. Varley, John Wiley & Sons, Chichester, 1983 and 1988, Chapter 1.

NUCLEAR SEXING FROM ORAL MUCOSAL SMEARS

Nuclear sexing stems from the discovery of sexual dimorphism in interphase nuclei by Barr and Bertram in 1949 (*Nature* **164**, p. 676). These workers identified in nerve cell nuclei from female cats a granule which stained deeply with certain dyes. No such granule could be seen in similar nuclei from male cats. The differentiating structure is called 'sex chromatin' or, after its discoverer, a 'Barr body'. The same sex dimorphism is evident in human tissues and is most easily seen in cells from the oral mucosa – the stratified epithelium lining the inside surface of the cheek.

Procedure

1. Thoroughly clean a slide and coverslip.
2. With the blunt end of a tongue depressor, scrape the inside surface of your cheek. Scrape vigorously so as to secure some of the deeper cells of this stratified epithelium. The very superficial cells are in poor cytological condition and are not satisfactory for demonstration of the sex chromatin.
3. Spread the desquamated cells and saliva on the surface of the clean slide.
4. Add a drop of 0.5% aceto-orcein to the smear.
5. Cover with a coverslip. Allow a few minutes for staining, then squash gently between sheets of filter paper.
6. Examine under oil immersion.

The sex chromatin-negative nucleus of a normal male has finely granular nucleoplasm, a well-defined nuclear membrane, and no sex chromatin body. A chromatin-positive nucleus from a normal female is similar but shows a small blob of sex chromatin lying hard up against the nuclear membrane (see Figure 6.1). In a few cases the sex chromatin body may lie in the centre of the nucleus, and in this position it cannot always be distinguished from a nucleolus.

There is a variable proportion of degenerating nuclei in oral mucosa smears, depending upon the depth within the membrane at which the cells have been removed, as explained above. These degenerating nuclei are coarsely granular or 'bubbly' in appearance and show no sex differentiation. Barr bodies are known to represent one of the X chromosomes of the normal human female complement. Two Barr bodies per nucleus in a female indicates the triple X (XXX) condition. One Barr body in a male indicates an XXY condition and Klinefelter's syndrome. No Barr bodies in a female indicates an XO condition and Turner's syndrome.

NUCLEAR SEXING FROM POLYMORPHONUCLEAR LEUCOCYTES (NEUTROPHILS)

Sex dimorphism can be demonstrated in the polymorphonuclear (PMN) leucocytes of humans. The nuclei from PMN leucocytes of normal females show a round body attached to one of the lobes of the nucleus by a thin stalk. This body, which is known as a 'drumstick' is absent from the PMN leucocytes of normal males.

Procedure

1. Thoroughly clean two slides.
2. You require to draw enough blood from yourself to make one or two good smears. Shake blood into your hand (or swing your arm about the shoulder to drive blood centrifugally) and prick the soft tip of the little finger with a sterile blood lancet. As soon as a drop of blood forms, dab it onto a clean slide and draw it out, using the end of another slide, into a smear. The more quickly you do this, the better will be the smear.
3. Dispose of the blood lancet in the 'burn bin' provided and put any blood-contaminated tissues in an autoclave bag. Put all unwanted blood-contaminated glassware into the sodium hypochlorite baths provided. Dispose of blood-contaminated products according to local regulations for health and safety.
4. Allow the slides to dry for about 2 minutes.
5. Place the slides in a Coplin jar filled with methanol. Leave to fix in methanol for 15 minutes.
6. Transfer the slides to a Coplin jar containing buffer solution. Add 3 ml of Giemsa stain and mix thoroughly by repeated pumping of fluid in and out of the pipette.
7. Leave for 10 minutes.
8. Flush the Coplin jar out with distilled water.
9. Remember that the slides must never pass through the surface of the Giemsa stain solution or the metallic surface film will make them very dirty and quite useless.

10. When all the stain has been washed out of the jar, remove the slides and dry them by vigorously waving them about in the air. Fast drying gives better uniformity of staining.
11. When the slides are completely dry, examine them without a coverslip using an oil immersion objective.

In an average female, one PMN leucocyte in ten will show a good drumstick. Critical microscopy is needed to identify these cells.

Draw exactly what you see in both oral mucosal and polymorphonuclear cells and do not forget to include a scale bar for all your drawings.

INFORMATION FOR INSTRUCTORS AND TECHNICAL STAFF

ALL STUDENTS MUST BE AWARE OF THE SAFETY PRECAUTIONS NECESSARY WHEN HANDLING HUMAN BLOOD AND UNDERSTAND THE REASONS FOR THESE PRECAUTIONS.

Equipment required to set up practical

Microbiological safety hood suitable for working with human blood
Waterbath or incubator at 37°C
Sterile Universal containers, screw-cap, Nunc, Cat. No. 3-64258
Safety, monovette Lithium heparinized syringe, Sarstedt Ltd,
    Cat. No. 02.1065.001
Monovette needles 20G × 1.5", Sarstedt Ltd, Cat. No. 86.1160
Rack to hold Universals.

Solutions required to set up cultures

McCoy's 5A Medium, Life Technologies Ltd, Cat. No. 041-06600
Fetal calf serum, Life Technologies Ltd, Cat. No. 011-06290
Phytohaemagglutinin (PHA), lyophilized, Life Technologies Ltd, Cat. No. 061-00576C or Wellcome Diagnostics Ltd, Cat. No. WRHA1580D
Sterile distilled water to reconstitute PHA
L-Glutamine, 200 mM, 20 ml, Life Technologies Ltd, Cat. No. 043-05030D
Penicillin/streptomycin solution, 5000 u/ml, Life Technologies Ltd, Cat. No. 043-05070.

If preferred, all the above items can be purchased as a tailor-made, pretested medium:

Chromosome Medium 1A with PHA, 100 ml, Life Technologies Ltd, Cat. No. 041-0167H
Colcemid, 10 µg/ml, Life Technologies Ltd, Cat. No. 061-0521C.

Preparation of cultures

Culture tubes may be set up in advance and stored overnight in the fridge. Cultures should be warmed to room temperature and inoculated with 0.4 ml of blood 70–72 hours before harvesting cells. One culture tube is sufficient for each student.

Working in a sterile hood and using **aseptic techniques** add the following to each 100-ml bottle of McCoy's 5A Medium (enough for 12 cultures):

12.5 ml fetal calf solution
2.5 ml reconstituted PHA
2.0 ml L-glutamine
2.00 ml penicillin/streptomycin.

Dispense approximately 9.5 ml into each Universal tube. Add 0.4 ml of whole blood to each tube, screw cap on tightly and incubate at 37°C in the dark. Shake gently twice a day to mix contents of tube. Check pH of cultures: medium should be peachy/pink, indicated by the phenol red in the medium. Add 0.1 ml of Colcemid to each culture, incubate for a further 1–1.5 hours, total time 70–72 hours. Harvest cultures quickly as extended antimitotic treatment condenses the chromosomes.

GENERAL EQUIPMENT REQUIRED IN LABORATORY

Bench-top centrifuge with 15-ml buckets to spin at 750–1000 r.p.m.
37°C waterbath
Slide drier
Disposable latex gloves
'Waste beakers' containing 5% sodium hypochlorite for discarded blood
  solutions
Autoclave bags to collect contaminated glassware and plastics separately
Bur 'N' bin for contaminated sharps.
Slide micrometers
England finder (grid reference slide for relocating preparations on any
  microscope)
Photomicroscope
Diamond pencil.

Equipment required per student

Photograph of human metaphase set
Thin white card, A4, scissors and glue pen
Two × 15-ml screw-cap, centrifuge tubes in rack or beaker
Disposable, 3-ml, plastic pipettes, Alpha Labs, Cat. No. LW4111

Glass Pasteur pipettes, siliconized using Repelcote, Merck Ltd, Cat. No.
    632164J
Jar of microscope slides in 70% ethanol
22 × 22 mm No. 1.5 coverglasses in 70% ethanol
Tissues
Filter paper
Compound microscope with oil immersion
Eyepiece graticule
Wooden tongue depressor, Richardsons, Cat. No. T2280
Blood lancet
Autoclave bags to collect blood contaminated glassware and plastics.

Solutions

5% sodium hypochlorite to mop up any blood spills
0.075 M potassium chloride at 37°C
Coplin jar of phosphate buffer, pH 6.8 (Sørensen)
3:1 methanol–acetic acid
Giemsa stain (0.5 g in 33 ml glycerol at 60°C overnight, add 33 ml methanol),
    Eastman Kodak, Cat. No. K1204387
XAM mounting medium, Merck Ltd, Cat. No. 36119, or equivalent
Immersion oil
0.5% aceto-orcein (1 g orcein in 45 ml acetic acid, heat to near boiling, add
    55 ml distilled water, filter when cool), Sigma Chemical Co., Cat. No. O7505
Coplin jar of methanol.

LABORATORY PRACTICAL 2: CELL SIZE, DNA CONTENT AND POLYPLOIDY

CELL SIZE AND NUCLEAR DNA CONTENT

Cell size and nuclear DNA content are related and are of considerable significance in relation to the duration of the cell cycle and the timing of development.

Amphibians (together with birds and reptiles) have nucleated erythrocytes, thus providing a population of single cells of uniform size and regular shape. Cell and nuclear volume and nuclear DNA content can therefore be measured quite accurately.

This practical is based on fresh heparinized blood from two species of amphibian that have widely different 'genome sizes' (amounts of DNA per haploid chromosome set).

Place a drop of the blood suspension on a clean slide and cover with a clean coverglass. Seal the preparation around the edges of your coverglass with nail varnish to prevent drying out. Then examine the preparation directly with and without phase contrast.

It is important to be sure of the shapes of the cells before trying to estimate their volumes. Take a freshly made unsealed preparation and gently move the coverglass while you are looking down the microscope: do this by pushing it with a dissecting needle. This will have the effect of moving the cells and rolling them over, so allowing you to ascertain their three-dimensional shapes.

Measure the cells in each blood sample using the calibrated eyepiece micrometer in your microscope. It may be necessary to use the high-power oil immersion objective.

Work out the cell and nuclear volumes (taking some account of the shapes of the cells) and write down your estimates of the ratio of the genome sizes of the three species, on the assumption that there is a linear relationship between volume and C-value. Give notes on the manner in which you arrived at these estimates, the assumptions you made and the likely sources of error.

THE SQUASH TECHNIQUE

The squash technique provides a simple, fast way of visualizing chromosomes and cell nuclei in animal and plant tissues. It is carried out in four main steps:

1. **fixation** of the tissue in ethanol–acetic acid (3:1);
2. **softening** and maceration of the tissue in 45% acetic acid;
3. **squashing** between slide and coverglass;
4. **making permanent**.

Fresh tissue is dissected from an animal and placed immediately in a freshly made up ice-cold mixture of 3:1. It can be left in this fixative for days – years – as long as it is kept below 4°C. The usual practice, if not making preparations immediately, is to place the tissue in 3:1 and keep it in the freezing compartment of a refrigerator. If the preparations are to be made immediately the tissue should be left in ice-cold 3:1 for a minimum of 15 minutes.

After fixation (and staining by procedures that will be described later) the tissue is softened by immersing it in 45% acetic acid. This has the effect of softening collagenous materials and causing the cells to swell and separate from one another. You can watch how opaque, newly fixed material becomes transparent and falls apart in 45% acetic acid. Of course, acetic acid is a fairly drastic treatment for any cell, and many of the cellular proteins are lost in this step. Amongst other things, the nucleolus tends to disappear in 45% acetic acid, and virtually all cytoplasmic structure is lost.

The softened tissue can easily be teased apart into what is as nearly as possible a single-cell suspension. This is normally done using about a cubic millimetre of tissue in a drop of 45% acetic acid on a slide. When the tissue has been teased apart, a coverglass is placed on top of the preparation, and excess liquid is blotted away by placing the slide in a fold of filter paper and blotting the area around the coverglass. You can then put your thumb on the region of the filter paper that overlies the coverglass, and press down as hard and evenly as you can. The trick of making good squashes is to press down without 'squidging' the coverglass sideways.

You then have a preparation that you can look at under a microscope. Remember however, that your preparation is *wet*. It should therefore be examined with phase contrast, unless it has been stained during the 45% acetic acid step. Moreover, it is not permanent and will dry out in due course and become useless.

At one time, making a squash preparation permanent presented some difficulty. Just think. You have squashed cells between slide and coverglass. To make the preparation permanent you must somehow separate the coverglass from the slide in such a way that the cells are not destroyed or lost, but remain attached to either the slide or the coverglass. The cells must then be dehydrated by immersion in ethanol and finally mounted in a transparent resinous substance that will preserve them and allow them to be examined by critical microscopy, and the condition of the slide must be such that it can be stored over a period of years and cleaned repeatedly.

Clearly, if you have a freshly made squash preparation, you cannot just take off the coverglass and expect to proceed with satisfaction. Try for yourself. The result is catastrophic!

In the early 1950s two men, Conger and Fairchild (1952) came up with the answer. When making a squash, use a slide that is thinly coated with a film of gelatin (a 'subbed' slide) and a coverglass that is thinly coated with a film of silicone (a 'siliconized' slide). After you have made your squash, put the slide on the surface of a block of solid carbon dioxide (dry ice) or immerse it in liquid nitrogen. Dry ice is much cheaper and easier to use. It effectively freezes the tissue to the sticky subbed surface of the slide. The tissue does not stick to the water-repellant surface of the coverglass. After everything has frozen hard (this takes about five minutes) you can quickly and simply flick off the coverglass with a razor blade, and if you have done the job properly then all your cells will be on the slide and the coverglass will be entirely clean. You must then place the slide immediately into a jar of 95% ethanol for a few minutes, remove it, put a drop of mounting resin over the preparation and cover with a clean, dry coverglass. The slide is then placed on a hotplate for a day or so to allow the resin to harden and set. Now you have a preparation that can be kept for ever.

POLYTENE CHROMOSOMES

The salivary gland cells of many dipteran larvae have polytene chromosomes. These chromosomes are amongst the largest known that are readily available for cytogenetic study.

The chromosomes owe their large size to the fact that they have replicated over and over again and the strands that represent the products of replication have remained close together, side by side, in perfect register with one another, rather like the strands of a rope.

More important than size are two other characteristics demonstrated by these chromosomes. First, the chromosomes reveal a distinctive pattern of transverse banding which consists of alternating chromatic and achromatic areas. These bands differ in thickness and in other structural features in a manner so specific that it permits the accurate mapping of each chromosome throughout its length. Second, the homologous chromosomes show a type of synapsis (somatic pairing) which appears as intimate as that characteristic of meiotic chromosomes in pachytene, such that each visible chromosome actually consists of the products of repeated replications of two homologous chromosomes. We may therefore expect to see the haploid number of polytene chromosomes in each diploid nucleus.

Certain regions of the polytene chromosomes undergo reversible changes in circumference during larval development. These 'puffs' represent areas of intense chromosomal synthesis of RNA. In *Chironomus*, the nucleolus organizer appears on one chromosome as an enormous puff. Puffs of this size are sometimes referred to as Balbiani rings (see Figure 7.4, Chapter 7). In general, a larva of a given stage of development will have very specific bands involved in puffs. Moreover, some puffs are tissue specific and some occur only in certain parts of the salivary gland.

The close somatic pairing of the polytene chromosomes allows the detection of structural differences between homologous chromosomes.

If, for instance, the following gene sequence, representing a simple inversion, were to pair:

Sequence 1   a b c d e f g h i

Sequence 2   a b c g f e d h i

then one would expect a pairing configuration like this:

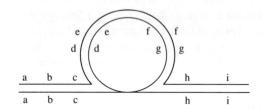

This is, in fact, precisely what is frequently encountered in polytene chromosomes from wild populations of *Chironomus*.

Removal of glands and fixation

Chironomus larvae are easy to collect and obtainable at all times of the year. *Chironomus tentans* has four pairs of chromosomes. In your preparations you will see only four chromosomes, because members of homologous pairs have fused with one another. If there are five chromosomes visible, then you have a preparation from *C. plumosus*, in which the shortest (fourth) chromosome does not pair.

Dissection

1. Select a large larva and place it on a piece of filter paper to drain off all the water. Transfer it to a clean slide, and place the slide on the stage of your dissecting microscope. Bring the larva into view under the lowest magnification. Identify the head and tail of the larva. The tail bears two prominent, stout appendages. The head lacks appendages.

2. Grasp the larva with a pair of forceps about half-way along its length. Then, using a sharp scalpel, cut cleanly across the rear of the first segment, so severing the head and first segment from the rest of the larva. The salivary glands will immediately be visible, oozing out from the cut end of the body. They are large translucent objects. Ask your instructor to show you if you are in doubt.

   Work quickly and do not let the isolated glands become dry.

3. As soon as you have recognized the glands, remove all other debris from the slide and add a few drops of 3:1 fixative. Also make fresh preparations of glands for examination in phase contrast as instructed below.

4. As soon as you have covered the glands with fixative, pick them up very carefully on the point of a needle or your fine forceps and transfer them to the small glass dish of fixative. You should collect glands from as many larvae as possible and put them all together in this vial. About five pairs of glands will ensure that you end up with a reasonable preparation.

Squashing

5. Set out in a neat row on a clean paper towel as many subbed slides and siliconized coverslips as you have glands.
6. Place each gland on a subbed slide in a few drops of 0.5% orcein in acetic acid.
7. Cover the slides with the lid of a Petri dish and wait for about 15 minutes. Add a little more orcein if necessary to prevent drying out.
8. Take a siliconized coverslip, dry it with a clean tissue, and then by wiping and blowing make absolutely sure that it is completely free from dirt and dust. Remember that one speck of dirt or one strand of cellulose from a tissue will undoubtedly be thick and tough enough to protect the salivary gland cells from effective squashing. Gently place the coverslip on the area of the slide occupied by the glands and aceto-orcein.
9. Take a clean filter paper and fold it in half. Place the slide in the fold of the filter paper, and fold the paper down so that you blot the preparation between the layers of filter paper. Apply *gentle* pressure through the filter paper to the coverslip and around the coverslip. You can then have a look at your slide with the microscope.
10. If it looks good (in the opinion of your instructor) then mark the slide with your initials using a diamond pencil, place it on the surface of dry ice and leave it there for at least 5 minutes. You can leave it on dry ice for as long as you wish.
11. Remove the slide from the dry ice and **immediately – without pausing at all – flick off the coverslip** with a razor blade and put the slide in the first Coplin jar of 95% ethanol. **You must do this quickly.** Keep the siliconized coverslip for re-use.
12. Leave the slide in the first jar of 95% ethanol for about 5 minutes and then place it in the second jar of 95% ethanol for a further 5 minutes and then in 100% ethanol for 5 minutes. Dip the slide gently into the jar of Histoclear or Xylene several times and then wipe it dry, except for the region around the preparation, which must on no account be allowed to dry.
13. Clean an ordinary coverslip thoroughly. Place one drop of XAM or other mounting medium on top of the preparation and gently lower the coverslip onto the XAM. Wait until the XAM has spread evenly under the coverslip, and then your preparation is complete. Be careful, however, for most mounting media take several days to set and dry, so do not accidentally 'squidge' the coverslip.

Phase-contrast observation of polytene chromosomes

Put a substantial drop of liquid paraffin on a very clean slide. Select a large larva and blot it dry on filter paper. Place the larva in the drop of paraffin and dissect out the glands in such a way that they are never exposed to air but remain at all times submerged in paraffin. Clear the debris from the slide and arrange the glands so that they are separated from one another, spread out, and lying under paraffin in a small pool of haemolymph. Then cover with a clean coverslip and compress very slightly. Now wipe away any excess paraffin from around the coverslip and examine your preparation in phase contrast. Many objects will be visible in fresh material that are lost in fixation and staining procedures. Identify the large nucleolus, the Balbiani rings on the smallest chromosomes, particles in the cytoplasm, the outline of the nucleus and the arrangement of the secretory cells.

Questions to ask and answer

- Where did the *Chironomus* come from?
- How many chromosomes can you see in each group?
- How long is the longest chromosome?
- About how many cross-bands are there per unit length of these chromosomes?
- How many cross-bands altogether?
- How many cells to a salivary gland?
- Do the chromosomes of small larvae differ in any way from those of large larvae?
- Identify, draw and explain the configurations of chromosomes that signify the presence of any major inversion heterozygosities.

INFORMATION FOR INSTRUCTORS AND TECHNICAL STAFF

Material required for nuclear size DNA content

Fresh, heparinized blood from two different species with widely different 'genome sizes' diluted in a suitable medium.

Any anuran compared with any urodele is suitable. *Xenopus* and *Notophthalmus* are ideal. One animal will suffice for a class of up to 50 students.

Place 10 ml of 0.1 M lightly buffered amphibian saline in a plastic vial or centrifuge tube.

Mix about 100 µl of heparin into the saline using a glass Pasteur pipette.
Save the same pipette for collection of blood.

Kill the animal by anaesthetizing in MS222, dissect to expose the heart, flood the region of the heart with heparinized saline and then nick the blood vessels immediately in front of the heart with fine scissors.

Collect blood into the tube of saline.

Make quite sure the animal is dead and dispose of its carcass in the proper way.

Distribute about 5 ml of blood suspension per four or five students.

Equipment required per student

Nail varnish (one bottle per five students)
Stage micrometer slide (one per five students)
Microscope slides and coverglasses
Immersion oil
Compound microscope with phase contrast and eyepiece micrometer in one eyepiece.

Material required for polytene chromosomes

Chironomus larvae, collected from local freshwater ponds, where they are found in the top layer of mud, especially near trees. It is easier to collect buckets of mud and sort in the laboratory. Layer some of the mud in a large white tray cover with water an inch deep and agitate a small section of the mud at a time; this brings the larvae to the surface. Some aquarist shops sell 'bloodworms' as live fishfood but for polytene studies these are far inferior to those collected freshly from the wild.

Equipment required per student

Dissecting microscope
Compound microscope with phase contrast
Two pairs fine forceps No. 5
10–12 subbed microscope slides
Subbing solution (0.1% gelatin, 0.01% chromic potassium sulphate)

> Dissolve the gelatin in hot water, cool, and add the chromic potassium sulphate dissolved in a small amount of water. Filter through a Millipore 0.45 μm filter. Sonicate racks of slides briefly in a weak detergent solution, rinse in running tap water then distilled water before dipping in freshly prepared subbing solution. Drain slides and bake for several hours or overnight at 65°C. Store in a dry place.

20–30 siliconized coverglasses in jar of 70% ethanol (re-usable)

> Dip coverglasses one at a time into a beaker of Repelcote. Lift out and drain off most of the Repelcote, blot on a tissue and store in 70% ethanol.

Razor blade, single edge
Petri dishes, 2 × 90 mm

Paper towel
Pipettes, small plastic disposable, Alpha Labs, Cat. No. LW4111
Filter paper, No. 1, 11 cm.

Solutions

Pulverized solid carbon dioxide in a polystyrene ice bucket (Dri-ice, cardice).
   A large, flat surface area is required for maximum contact in order to freeze slides quickly.
45% acetic acid
3:1 ethanol–acetic acid
Two Coplin jars of 95% ethanol
One Coplin jar of 100% ethanol
Coplin jar of Histoclear (non-toxic xylene substitute)/ National Diagnostics, Cat. No. HS200
Liquid paraffin
0.5% aceto-orcein
Small vial of XAM or other equivalent histological mounting medium.

LABORATORY PRACTICAL 3: MEIOSIS

If at all possible, students should be provided with prepared slides of stained histological sections of testis material from an amphibian, an insect or a mammal. These should be used to give a general impression of the histology of testes, their shapes and the sizes of their cells and chromosomes.

For squash preparations use testes from locusts. This choice of material is based on three considerations. First, the animals are easy to obtain and maintain in the laboratory and they are not endangered species. Second, these animals have quite large chromosomes and they offer good opportunities for the study of chromosomes at all stages of the meiotic divisions. Third, meiosis in locusts shows most of the features that have been described for meiotic divisions. Essentially, the chromosomes do what 'the book' says they should do, and a few more interesting things besides.

In order to visualize the material in your squash preparations, you will use a staining technique that was devised in 1924 by Feulgen and Rossenbeck, generally referred to as the Feulgen reaction. This staining technique when properly carried out is absolutely specific for DNA; nothing else stains. It is an extremely valuable technique for the cytologist, and is widely applied today in quantitative studies of amounts of DNA in cell nuclei and chromosomes. Its basis is explained in Chapter 2 and also in Chapter 10 of H.C. Macgregor and J.M. Varley, *Working with Animal Chromosomes,* 2nd edn, 1988.

The Feulgen reaction involves the use of Schiff's reagent, which is colourless until it reacts with aldehyde groups, whereupon it turns bright pink.

IF YOU ARE SLOPPY WITH THE FEULGEN REACTION YOU WILL END THE DAY WITH PINK FINGERS, FACE, CLOTHES, PINK EVERYTHING! IT WILL WASH OFF YOUR SKIN WITHIN A DAY OR TWO, BUT IT IS DIFFICULT TO REMOVE FROM CLOTHING.

LOCUSTS

Locusts are a convenient source of material for studies of meiosis, since they can be reared in the laboratory at any time of the year, they have chromosomes that are relatively large and easily studied, and they show a number of interesting cytological features.

You are provided with fifth instar males of the desert locust *Schistocerca gregaria*. At this stage of development, the testes contain cells in most stages of meiosis, although there are unlikely to be many in the late spermatid stage.

Behead the locust and pin it on its front in the wax-filled dissecting dish. Make a single longitudinal median cut through the dorsal wall of the abdomen.

1. **Dissect dry**. Use the binocular dissecting microscope with the lowest magnification and a thin pencil beam of incident light. Immediately underlying the body wall you will find a yellowish mass of tissue which lies in the midline of the body. This is the testicular material and associated fat. Remove this material carefully using fine forceps. It appears as a single mass, but in fact comprises two testes which lie close together, ensheathed in common connective tissue and fat.

2. Place all the material in 3:1 in a solid watchglass and, as it is fixing, tease it apart gently with forceps and a needle. You will see that each testis is a bundle of small tubes rather like a bunch of bananas. Each of these tubes or **ampullae** will show a complete range of stages of meiosis, with gonial divisions at the top (anterior) of the tube and spermatids at the bottom (posterior).

3. After your material has been in fixative for at least 15 minutes, transfer the pieces of testis into the vial containing 5 N hydrochloric acid, and leave for **precisely 20 minutes**.

4. Drain off the 5 N hydrochloric acid with a Pasteur pipette. Fill the vial with distilled water. Drain off all the water and refill the vial with distilled water. Repeat this washing procedure. The aim is to get rid of all traces of the hydrochloric acid before adding the Schiff's reagent.

5. Half-fill the vial with Schiff's reagent, put on the cap and leave for at least an hour. You will notice that the pieces of tissue begin to turn pink almost immediately. During this time you should proceed with orcein-stained squash preparations of locust testes (see below).

6. Drain off all the Schiff's reagent with a Pasteur pipette. Fill the vial with SO_2 water. Drain it off and refill it with SO_2 water and leave for 10 minutes. Treatment with SO_2 water bleaches out all dye that is not bound to DNA.

7. Clean off a series of subbed slides. Make sure they are absolutely free from dust and grit. Put two or three ampullae in a small pool of 45% acetic acid on the slide. Using a needle and a pair of very fine forceps, gently separate the ampullae from one another so that there is a space of at least 2 mm between them.

8. Still using the needle and forceps, break up each ampulla into the finest possible pieces, until you are left with as many patches of disaggregated testis as you had ampullae in the first place. Now use the fine pair of forceps and carefully pick out any large pieces of connective tissue or other debris, leaving only a relatively homogeneous sludge of fine tissue fragments and dissociated cells.

9. Carefully clean a siliconized coverslip, making sure there are no pieces of fluff or grit on it. Drop the coverslip on top of the pool of stain and minced tissue. Blot the preparation lightly with a filter paper, taking great care not to 'squidge' the coverslip sideways. Then squash firmly but lightly. Have a look, using low-power phase contrast. See if you can find areas where there are lots of diplotenes and first metaphases. Are the chromosomes of each individual group spread out and dissociated from one another, or are they still bunched

up together? If the latter, then take the slide off the microscope and squash it again, only this time with a little more pressure. Good preparations should then be made permanent in the usual way by freezing on dry ice, dehydrating in ethanol and then mounting in resin under a coverglass.

To make orcein-stained preparations, place four ampullae on subbed slides. Put several drops of aceto-orcein onto each slide so that the tissue is covered. **Do not let the preparations dry out**. Put your slides in the Petri dishes provided and leave them for about 15 minutes. Take the slide out of the Petri dish and place it on a clean filter paper on the bench. Remove excess stain with a Pasteur pipette or a piece of filter paper, so that your tissue is surrounded by a pool of stain not more than 1 cm across. Then tease the tissue apart, squash and make permanent as described above.

Questions to ask

- What species of locust are you working with?
- What is the haploid chromosome number?
- How would you describe the karyotype?
- What is the chiasma frequency?
- What distinguishes the X chromosome at diplotene?
- Which is the heterogametic sex in this species?

NEWTS AND SALAMANDERS

Newts and salamanders are tailed amphibians belonging to the subclass Urodela. All urodeles have large genomes, meaning large amounts of DNA per haploid chromosome set, and most have relatively small chromosome numbers (n = 11 to 14). The crested newt, *Triturus cristatus* has 21×10^{-12} g of DNA per haploid chromosome set, which is roughly seven times as much as man. It has a haploid chromosome number of 12, as compared with 23 in man. The North American newt, *Notophthalmus viridescens*, has 35×10^{-12} g of DNA per haploid set of 11 chromosomes.

There are urodele amphibians that have much larger genomes but about the same haploid chromosome number. One species of salamander (*Bolitoglossa subpalmata*) that is found in Central America, near Panama, has about 70×10^{-12} g of DNA embodied in 13 chromosomes, and it can claim to be the animal with the 'biggest chromosomes in the world'. This, then, is why urodeles are good for chromosome studies. In cytology, seeing is the basis of most experiments and discoveries, and the bigger a thing is the easier it is to see. Urodeles are relatively common, easy to keep in the laboratory, and they have very big chromosomes. Besides this, they provide us with several very good sources of chromosome material. At certain times of the year, in temperate species, their testes have spermatocytes that are in all stages of the meiotic process. In species from tropical regions, spermatocytes in *all* stages of meiosis are present throughout the year. Amphibians also provide good sources of other chromosomal material. Their stomach and intestine have epithelia that are rich in dividing cells and they therefore provide an excellent source of **mitotic** chromosomes. Their ovaries consist of oocytes that are in all stages of oogenesis, and these provide superlative material for the study of chromosomes in the **lampbrush** condition that is so characteristic of developing ovarian eggs in all vertebrates other than mammals.

If you are fortunate enough to have access to a plentiful supply of adult male amphibians that have meiotically active testes, then it is recommended that you obtain a copy of the second edition of *Working with Animal Chromosomes* and follow precisely the protocol given in that book for studying meiosis in male newts and salamanders.

Questions to ask

- What species of amphibian are you working with?
- What is the diploid chromosome number for this species?
- How would you describe the karyotype in general terms: range of chromosome size, number of metacentrics, submetacentrics, sex chromosomes, etc.?
- What is the chiasma frequency (mean number of chiasmata per chromosome)? Determine it for at least one named species.
- How are the chiasmata distributed? Does any one chromosome seem strikingly different from the others in this respect?
- Are there any evident heteromorphic chromosomes (chromosomes that do not pair up)? If you think there are, then draw an example.
- Are there any chromocentres in meiotic prophase and, if so, how many? Draw an example.
- How do the chromosomes of this amphibian compare in size with those of the locust and with human chromosomes?

INFORMATION FOR INSTRUCTORS AND TECHNICAL STAFF

Material required

Prepared salamander testis slides

Salamander testes, collected in advance and fixed in 3:1 ethanol–acetic acid

Locust testes from late fourth or fifth instar male locusts, *Schistocerca gregaria*. Can be collected over a period of several months before the practical, fixed in 3:1 and stored at $-20°C$

Locusts, Philip Harris Education Ltd, Cat. No. M45343/9.

Equipment required per student

Dissecting microscope
Compound microscope
Two pairs fine forceps No 5.
Two small glass or plastic vials
Subbed slides
Siliconized coverglasses
Two Coplin jars of 95% ethanol
Coplin jar of 100% ethanol
Coplin jar of Histoclear
Razor blade, single edge
Two Petri dishes, 90 mm
Two embryological watchglasses (solid glass base)
Small dissection dish and a few dressmaker's pins
Nylon tapper or pencil with a rubber on the end
Tissues
Small vial of XAM or equivalent histological mounting medium.

Solutions

Pulverized solid carbon dioxide in a polystyrene ice bucket
5 M hydrochloric acid
45% acetic acid
SO_2 water, harmful vapour. Prepare just before use
0.5 g sodium metabisulphite in 190 ml distilled water, add 10 ml 1 M
    hydrochloric acid
Schiff's reagent, harmful vapour

Dissolve 1 g of basic fuchsin in 22 ml of hot distilled water. Add 20 ml of 1 M hydrochloric acid, mix, then 2 g of sodium metabisulphite. Leave in the dark overnight. Working in a fume hood, filter solution through a Whatman No. 1 filter paper using a Buchner funnel. It is important that the filtering apparatus is absolutely dry as any contact with water will cause the solution to go pink. Store the colourless Schiff's reagent in a clean, dry, dark bottle in the fridge. (Basic fuchsin Eastman Kodak, Phase Separations Ltd, Cat. No. C1762.)

Distilled water

3:1 ethanol–acetic acid.

LABORATORY PRACTICAL 4: LAMPBRUSH CHROMOSOMES

The largest known chromosomes are found in the developing oocytes of various vertebrates. These so-called 'lampbrush' chromosomes are in many ways the most favourable of all for experimental manipulation. Despite the fact that techniques for their isolation from living cells were worked out nearly 50 years ago, very few workers have made use of this remarkable material. Part of the reason for the lack of interest seems to be the mistaken notion that these chromosomes are difficult to handle or require elaborate equipment. Quite the opposite is true, as we hope you will see during this laboratory period.

Lampbrush chromosomes are now known to be present in the developing oocytes of all vertebrates (except mammals) and most invertebrates. The ease with which the chromosomes may be studied varies widely from group to group, and even between members of a single family. The consistency of the nuclear sap is an important variable; the sap must be sufficiently gelatinous to permit certain manipulations, yet not so rigid that it fails to disperse after the nuclear membrane has been removed. This protocol is based on the use of material from the North American newt, *Notophthalmus viridescens,* and the frog, *Xenopus laevis.*

You are provided with two square embryo cups containing pieces of ovary from *X. laevis* and *N. viridescens* freshly removed from the animals.

ISOLATION OF THE NUCLEUS

1. Place a small bit of ovary in 5:1 solution. This is a mixture of five parts 0.1 M potassium chloride and one part 0.1 M sodium chloride, and contains the two cations in the proportion existing within the oocyte nucleus. The solution may be lightly buffered to pH 6.8–7.2 with phosphate at a concentration of not more than 0.005 M, if desired. Note that the smallest oocytes are nearly transparent. At about 0.5 mm diameter, yolk accumulation begins and the oocytes become cloudy. Those over about 0.8 mm diameter are completely opaque.
2. Each oocyte is covered with a thin layer of follicle cells and prominent blood vessels. It is not necessary to remove the oocyte from the follicular membrane. Puncture the cell with the forceps or a needle and squeeze gently. The nucleus appears as a clear 'bubble' embedded in a ribbon of yolk flowing from the hole. In some instances the nucleus may be almost clean of yolk at this stage; at other times it is fairly heavily encrusted.
3. Using a pipette with a mouth of between 0.5 and 1 mm internal diameter, suck the nucleus in and out several times to remove the adherent yolk. The pipette should be filled with 5:1 solution before starting the cleaning

procedure and care should be taken not to include an air bubble in the pipette. The nucleus is really quite sturdy and may be bounced off the bottom of the dish in order to remove bits of yolk. However, it should not be allowed to settle on the glass, as it may adhere firmly and tear when sucked loose.

The nucleus will begin to swell immediately after isolation. This swelling serves a useful purpose in separating the nuclear membrane from the underlying chromosomes.

Eventually, however, the nuclear sap will become fluid and the chromosomes will sink to one side of the nucleus. Therefore work as rapidly as possible once the oocyte has been opened.

ISOLATION OF THE CHROMOSOMES AND NUCLEOLI

4. The chromosome isolation can be done under a magnification of 20–40× depending on personal preference. Illumination is critical. Most workers prefer a strong lateral illumination with no transmitted light. You are provided with an **opaque black** stage plate for your binocular. Arrange your light to focus sharply to the smallest possible spot, pointing towards you and shining on the centre of the microscope stage at an angle of about 15°.

5. As soon as the yolk has been removed from the surface of the nucleus, transfer the nucleus to a flat-bottomed well slide, previously filled with chromo-some isolation medium. The simplest and usually the most effective medium consists of five parts of 0.1 M potassium chloride to one part of 0.1 M sodium chloride with formaldehyde added to a final concentration of 0.5%. The well of the slide should be completely filled so that the liquid has a slightly convex surface.

6. Grasp the top of the nucleus with one pair of jeweller's forceps, taking care to secure a good grip while not actually rupturing the membrane. Lift the nucleus just clear of the bottom of the chamber, and then very carefully insert the point of a tungsten needle just under the surface of the nuclear membrane a short distance away from the points of the forceps. Now tear the nuclear membrane by moving the point of the needle down and around the nucleus. If this is done quickly and neatly the contents of the nucleus

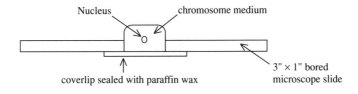

Nucleus   chromosome medium

coverlip sealed with paraffin wax

3" × 1" bored microscope slide

will spill out onto the bottom of the chamber in a single gelatinous lump, and they will then disperse slowly. **The secret of success is being fully prepared, and then working rapidly and carefully.** It is particularly important not to dig deeply into the nucleus with your forceps and needle.

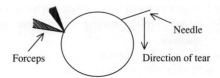

Forceps   Needle

Direction of tear

If at any time the nuclear contents begin to extrude spontaneously through the small hole in the nucleus, one should immediately abandon the preparation and begin with a fresh nucleus. Only fragmented chromosomes will be found in such preparations. The ease of the isolation depends a great deal on the 'stiffness' of the nuclear contents; the stiffer the nuclear sap, the easier the dissection. Certain species have naturally stiff sap, in some cases so gelatinous that the nuclear membrane can be peeled off piece by piece without disrupting the contents.

A coverslip should now be added. Hold a coverslip about 5 mm above the depression and drop it into place. It must be dropped so that the surface of the coverslip is parallel to the surface of the slide. If not, surface tension effects will destroy the preparation.

Alternative methods of preparation can be found in H.C. Macgregor and J.M. Varley, *Working with Animal Chromosomes*, 2nd edn, 1988.

OBSERVATION

The chromosomes will eventually spread out evenly on the bottom of the well-slide chamber. Scattered among them will be several hundred nucleoli. In addition, there may be a very large number of smaller granules, particularly if the nucleus came from a large oocyte. Dispersal of the sap and settling of the chromosomes may take as little as 5 minutes or may extend over a period of several hours. As already mentioned, much depends on the species, the size of the oocyte and the general condition of the animal from which the oocytes were removed.

Critical observation at high magnification requires the use of an inverted microscope since the chromosomes lie on the bottom coverslip of the well chamber. Phase-contrast optics are essential. If an inverted microscope is not available, low-power observations may be made through the liquid from above.

- Record the diameters of oocytes from which you isolate nuclei, the diameters of nuclei, the appearance of the dispersed nuclear contents with careful attention to scale and size, and the average lengths of the lateral loops that project from the chromosomes' axes.
- If there are nucleoli amongst the chromosomes in your preparation, count them – how many?

Instructors and students are strongly advised to read Chapter 6 of *Working with Animal Chromosomes* in order to obtain a fuller understanding of the reasons behind various aspects of this rather special protocol.

INFORMATION FOR INSTRUCTORS AND TECHNICAL STAFF

Material required

Oocytes from any of the following: *Xenopus laevis, Triturus, Notophthalmus, Pleurodeles.*

Kill the animal by anaesthetizing in MS222. Open the abdomen and remove all the ovary. Distribute small pieces of ovary into dry, clean embryo dishes and seal tops onto the dishes with petroleum jelly. Place on ice. Usually, one small newt will provide enough for four or five students, one large newt will suffice for ten, and one large *Xenopus* will provide for an entire class of 20–40 students.

Make quite sure that the animals are completely dead before disposing of them.

Advance preparation

Fine watchmaker's forceps. Hone the tips with an oiled Arkansas stone and a piece of very fine waterproof abrasive paper to make sure that they meet exactly. Stainless steel forceps are less suitable for this purpose: they tend to bend and distort too easily. Examine the points under the dissecting binocular.

Narrow-bore Pasteur pipettes. These are best prepared from long-form Pasteur pipettes. Position the middle of the wide barrel of the pipette over a hot, narrow Bunsen flame and rotate it continuously until the heated region is completely flexible. Then swiftly and smoothly pull the pipette out to a distance of arm's length and break it off leaving a long stem. Square off the end of the pipette about 5 cm from the shoulder by stroking the stem with a diamond pencil and snapping off cleanly at the resulting scratch. The tip of the pipette should be between 0.5 and 1.0 mm internal diameter. Touch the tip of the pipette very briefly to the side of the Bunsen flame to round off the sharp edges. Inspect the pipette under the binocular dissecting microscope to see that it has the right qualities.

Tungsten wire needles. Fuse a 3-cm piece of 0.35–0.40 mm thickness tungsten wire into a piece of Pyrex tube about the size of a pencil so that about 2 cm of the wire is projecting from the end of the glass holder. To prepare a point on the wire, molten sodium nitrite is used. Half fill a nickel crucible with solid sodium nitrite, stand in a pipe clay triangle over a Bunsen burner flame and melt the sodium nitrite gradually. Dip a tungsten needle into the molten sodium nitrite (be prepared for it to 'spit'). If the nitrite is hot enough the wire will immediately become incandescent. Dip the tip in several times for a few seconds, wash the needle in running water and inspect the point under the binocular microscope. The point should be very sharp and clean. Flame the needle a few times in a Bunsen burner flame.

WEAR PROTECTIVE SPECTACLES AND GLOVES FOR THIS OPERATION.

Lampbrush observation chambers. There are two ways that these chambers can be made:

1. The best chambers are made using bored slides. This requires the skill of a glassblower to drill 5-mm holes through microscope slides with either a diamond or ultrasonic drill. If the equipment and expertise are available this is the quickest method for producing hundreds of re-usable slides, as it is possible to drill a stack of five or six slides together. To prepare lampbrush slide chambers, first clean the bored slides thoroughly. With a heated copper or brass rod 1–2 mm thick put two small dabs of paraffin wax on either side of the hole, each about 2 mm from the edge of the hole. Thoroughly clean an 18 mm or 22 mm No. 1.5 coverglass from 70% ethanol and place one over the hole in each slide. Now wave a very small Bunsen burner flame over each coverglass until the wax has melted. Take care not to crack the coverglass or slide by excessive heating. If too much wax has been used, it will ride up the walls of the chamber and it will be impossible to fill the chamber with liquid thereafter.
2. Cut a square piece of double-sided sticky tape about 30 mm^2. Then cut a piece of 0.5 mm Teflon plastic sheeting the same size. Stick the tape firmly and evenly to one side of the plastic square. Using an ordinary office ring-binder hole-punch, make a hole in the centre of the plastic square. The hole must be clean and round. Remove the remaining backing sheet and stick the plastic square firmly and evenly to the middle of a clean microscope slide, so forming a chamber. Preparations made in these chambers should be covered with coverglasses that are smaller than the plastic square.

Equipment required per student

A binocular dissecting microscope (a simple kind with a fixed magnification at around 20× or a variable zoom magnification between 10× and 40×). It should be fitted with a calibrated micrometer eyepiece, a black stage plate and have an associated incident light source that can be focused to a sharp intense beam shining on to the middle of the stage plate at a low angle.

A mounted needle with a moderately sharp but slightly rounded point.

Two pairs of fine forceps, one No. 4 and one No. 5, the latter with very sharp points that meet perfectly.

A tungsten wire needle with a perfectly clean and very sharp point.

A narrow-bore Pasteur pipette with a long thin barrel ending in a mouth of about 0.5 mm diameter.

Several ordinary disposable Pasteur pipettes.

Lampbrush observation chambers, a minimum of three per student.

Coverglasses in 70% ethanol.

Three or four embryological watchglasses.

About 250 ml of 5:1 0.1 M KCl/NaCl (preferably autoclaved and cooled to room temperature).

About 50 ml of 5:1 0.1 M KCl/NaCl plus 0.5% formaldehyde.

A square plastic Petri dish with lid, containing a filter paper moistened with 0.1 M KCl/NaCl and a piece of glass rod bent into a U to serve as a bridge for supporting up to three microscope slides clear of the wet bottom of the dish. This serves as a moist chamber for storing preparations while they are awaiting examination.

An empty 500-ml plastic beaker into which waste isolation medium can be tipped. (This exercise generates no toxic waste.)

A small Petri dish or a square embryo dish containing a small portion of freshly removed ovary (about 20 large oocytes and associated small ones), sealed with a well-fitting lid and placed on wet ice. The ovary must not be placed in any kind of liquid medium at this stage, but kept cool and moist in a small sealed dish.

Species index

Index